I0493488

COSMOLOGY

REVISITED

Or

An Audacious Attempt to

Produce a Workable

Theory of Everything

By

GWH

@2012
G.W.Harper
Arlington, WA

DEDICATED TO

KATHY

For nearly sixty years of

Love and patient endurance

AND

GEORGE, CELIA AND JOHN

For turning into fine people

Despite my own failings

"It ain't what you don't know that's so apt

To hurt you as the things

You do know that

Ain't so!

Josh Ward; American humorist ca. 1845

{We could learn from him}

FORWARD

WHAT THIS ISN'T

World War II was not only the most destructive war ever fought; it was also the most productive of ever fought; it was also the most productive of technology across the entire spectrum of science, from medicine to A- and H-bombs, radar, computers and jet aircraft to rocketry and the increasingly accurate understanding of the universe. As our tools have sharpened the techniques of astrophysics have flowered at a truly remarkable degree. None of these technological advances are questioned here.

Neither is this a recapitulation of the increasing sophistication of our machinery when it comes to the extraction and presentation of data. I accept all of these uncritically, despite the inevitability of occasional errors; errors which are usually soon corrected on review.

I believe this is all to the good, but I can perceive a number of fundamental flaws to be reckoned with; several of which have persisted unremarked for centuries. A brief visitation focuses on these flaws and the consequences they may have had on our theories. This revisitation is one of the objects of my cosmology here.   But there is quite a bit more to be said on the matter and not everything is sweetness and light.

WHAT THIS IS

Sad to relate, there is one science which has never existed; and this lack has subtly infected the whole science of astrophysics.

This 'forbidden' science is the science of studying scientists, not as scientists *per se* but as men and women motivated by every passion available to ordinary mortals. They love. They hate. They seek fame and have a carefully concealed contempt for the money grubbers who seek no more of life than the acquisition of little pieces of green paper. They are arrogant pups who bask in the glory of their position and proudly display their diplomas and trophies for admiring students to gape at. Some are shy and virtually reclusive. Others seek to be the head of every parade.

Some devote their lives to an underground neutrino detection 'telescope' while others cheerfully man radio telescopes. More than a few have never looked through a telescope but instead have devoted their lives to teaching students in schools and universities.

Many are blindly traditional and tend to reject any data which conform to their system of thought. Others are subtly motivated by the dogmas of their theology and persist in casting every new finding into the porridge pot of their faith. In short, they are all members of the human species and subject to all the flaws which beset us.

One consequence of this is a deep-seated desire to propose arcane postulates which often border dangerously close to pure mysticism while at the other extreme they tend to ignore certain ideas because they are too disruptive to be convenient.

An example of this may be found in the famous 'red shift' as a measure of time and distance. The wave length is shifted toward the red end of the spectrum being emitted by an object which is receding from you and to the blue end of the spectrum of an object coming toward you. The degree of shift is a function of the speed of recession or approach.

This was the Holy Grail of the astronomical fraternity! At last, they had a measuring stick for the universe. They could measure time and distance with unprescented accuracy and were on the verge of a definitive model of everything

Unfortunately, there were a few recalcitrant souls lurking in the bush. One of them, I do not recall his name, possibly because he was placed in purdah for his temerity, advanced the idea that possibly light grew fatigued over its long journey and thus at least a portion of the red shifting was due to light fatigue and the shifting at extreme distances might not as perfect a measuring stick as advertised.

The response to this heresy was both quick and entirely predictable. "How dare he? Just when we get our measuring stick he steps in and seeks to demolish it!" On this note it was cast into oblivion, where it has remained ever since. But it is not demolished, only unproved... as is the red shifting model currently in place.

What we see here is not a case of who is right and who is wrong. It is a case of selectively choosing convenience by ignoring inconvenient ideas whenever they threaten existing dogma.

There is also a particularly dangerous tendency toward a myopia which forecloses gestalt thinking. Since the example of Uranus will not be cited until Book 3, I will mention it here to illustrate my meaning.

The polar axis of all save one of the system's nine inner planets is roughly the same as the polar axis of the sun. Added to this, the direction of rotation conforms to the solar norm... again except for Uranus.

What makes Uranus different is the fact that its polar axis is tilted a full 98 degrees toward the sun. This led to a wholly understandable wish to figure out how it happened. And here we have a glaring instance of myopia.

A pair of competing explanations was soon forthcoming. The first invoked the existence of a rogue Jovian sized planet in a wildly eccentric orbit which chanced to cross the orbit of Uranus at the same time Uranus arrived at that point in its orbit. Tidal forces forced Uranus to tilt in response to the majesty of this rogue planet; which was given the appropriate name of 'Nemesis'. The second, and more prosaic, theory proposes that the magnetic core of Uranus decided to roll over and the magnetic currents rising from the interior dragged the rest of the planet into its odd inclination.

Both theories are wildly improbable by reason of their myopia. It is easy to see what happened. Both theories focus exclusively on Uranus and ignore the simple little fact that Uranus also boasts five Jovian type satellites along with a partial ring which falls within the Roche limit!

None of these save Miranda (the innermost) has an orbital eccentricity greater than 0.005. A Nemesis passing through close enough to tilt an entire planet is going to play hobb with the orbits of any satellites which happen to be present. A shift in the magnetic axis at the core of Uranus is also certain to distort orbits. In short, the condition of the Uranian satellite orbits precludes both 'solutions'. *Sic semper myopia*!

An entirely different problem arises when we consider the rigidity of human conventions. Take the simple equation 1+1=2.

Who could possibly question it?

Actually there is a subtle ambiguity here. Take a 1 and append another 1 and you arrive at 11!

Matters worsen when you go to 111. If you are a traditional Roman this equals 3! But it can also be interpreted as 11+1=12 or an entirely different number if you are playing around in binary or octal!

In short we may confuse ourselves by subtle biases within our own minds which persist despite our belief that we are being totally objective.

These merely illustrate two sorts of errors which pass unremarked by the astrophysical fraternity. More exist which will be discussed in due time but these will suffice for now.

WHAT OUR OBJECTIVE IS

I have two major goals here. The first requires a survey of the history of astronomy from the earliest days in order to purge the record of those errors of mythology and methodology which have persisted as undercurrents to our thinking for long aeons until the present.

The second is to attempt to reconstruct cosmology in a form which is free of these myths and biases and is based on a purely logical interpretation of observations which is absent mystical postulations needed to support the myriad conclusions forced on us in our efforts to sustain them.

On this note we can proceed.

CHAPTER 1
IN THE BEGINNING

Much of current cosmology is philosophical rather than scientific. Physical observations are interpreted in terms of existing perspectives, not all of which are proven. And even when there is evidence to the contrary mistaken conclusions are adhered to because of popular beliefs. A pair of ancient examples of this habit provides an introduction to an otherwise cloudy argument.

Contrary to popular belief, when Columbus set sail from Spain in 1492 he was not alone in believing that the world was spherical. Centuries earlier the Babylonian astronomer Nabu Rimanni, (ca. 750 b.c.e.) obviously took it for granted that the Earth was a globe. Around 400 b.c.e., another Babylonian astronomer, Kidinnu (known to the Greeks as 'Kidinnu') in his canon of eclipses, cited eclipses which would not be visible to the Babylonians because they "would be seen only on the other side of the world"!

Add to this the fact that when Columbus sailed the ocean blue virtually every educated man in Europe was fully aware that the Earth was spherical. The Phoenicians and the Greeks had both correctly interpreted the fact that ships sailing off in the horizon disappear from the bottom up as proving that the Earth is a globe, and this knowledge had never quite been lost even in Europe. Columbus therefore introduced no new theory into the argument. In fact, what he did introduce was one whopper of a mistake!

Apparently somewhere along the line an unknown geographer disagreed with the rest of the mapmakers and mathematicians and decided that the circumference of a circle is pi times the radius rather than pi times the diameter; thus reducing Earth's circumference from 24,000+ miles down to 12,000+.

Columbus evidently stumbled across a map which mirrored this error. Given that the diameter of the Earth is 8,000 miles while Columbus believed it to be only 4,000, he calculated that the East Indies were only a bit more than 6,000 miles distant if he sailed west versus around 10,000 if he took the conventional route (which began with a 4,000 mile sail south to the Cape of Good Hope followed by an approximately like distance north after rounding the Cape). By sheer coincidence Columbus chanced onto the Island of Hispaniola after sailing almost precisely the distance he calculated would take him to Java in the East Indies.

Accordingly, he called the natives "Indians" and the islands the "West Indies," both of which names adhere to them to this day!

Nor is this the end of my little dissertation anent Columbus. He was not the man who discovered the new world! The Chukchi of eastern Siberia (or their predecessors) had ongoing contact with the Inupiat peoples of North America for more than 6,000 years before Columbus ever set sail. Archaeologists have discovered the remains of a Viking longhouse and settlement dating to around 950 a.d. in Labrador. It is possible (but unlikely) that the Irish Saint Brendan visited North America a few centuries before Columbus. An abrupt change in pottery styles, occurring about 400 a.d. in Peru gives evidence of a Joman (Japanese) presence, most likely a dismasted fishing smack caught in a transoceanic current and grounding in South America, leaving its crew with no alternative save to join the Indians, intermarry, and pass on some of their genes and pottery designs for future generations.

There is the Mesa Verde archaeological dig in Chile which antedates the Folsom People of North America, plus still suspect evidence of Phoenician presence in the Amazon River delta of Brazil and apparently antique fire pits further inland. And of course, there are the misnamed American Indians to attest to dozens, if not hundreds of 'discoveries' before Columbus. (And, as an aside, the Indians never knew they were lost!)!

So we have a badly flawed theory combined with an absurd claim to his 'discovery'. Given all this one might ask, why all the hooplah? What are we really celebrating?

Answering my question; Columbus still deserves respect and admiration from all of us. He had a theory and he pursued it for years until he was finally able to 'prove' it by the best evidence available at the time. The fact that more than five centuries later the rationale is regarded as laughable does not matter in the least. He dared and he did. Even more to the point, alone of all the others of European origins he made it stick. His voyage approximately doubled the size of our world... which was no mean achievement for a solitary man. We are right to celebrate what he did then, if not his motives.

Skip along to Copernicus and we find another group of anomalies. Here we have a high ranking Roman Catholic Bishop who enjoyed dabbling in astronomy. Being a major prelate he was highly literate and an insider in the political and theological infighting of the era. Doubtlessly he foresaw the uproar his model universe, where the Sun rather than Earth is the center of the universe, would create. Being a sensible man, he printed a few copies of his work and left instructions that they were to be distributed only after his death.

True to his expectations, a major brouhaha was stirred up. But I suspect it was not the one he anticipated!

Being a churchman and knowing the Church history of obsessive orthodoxy and firm conviction it already knew everything worth knowing, he no doubt anticipated a major furor within the Church. And in truth, there was initial unrest, but a little added caution by Copernicus generally muffled it. As an additional way of thwarting official dogmatic criticism, he carefully explained that he did not claim that the Earth moved around the Sun only that for purposes of predicting the motions of planets and ascertaining the positions of astronomical bodies it would be more convenient to treat the Sun rather than Earth as the center of the solar system.

This largely left the Church with little to bite on. He had died in good order and there was no basis for exhuming him and burning his corpse at the stake, as was done on occasion. Nor was he claiming that the Sun was the center of the solar system; merely that for certain purposes of mathematics it simplified the calculations. Given this, the Church held back. Unlike the somewhat laterGalileo, He was not officially declared a heretic and buried in unsanctified ground absent any of the usual monuments accorded upper level churchmen, but he was planted off in a remote corner of the graveyard, which meant he was a not too highly regarded suspect. There he remained for nearly 500 years before finally being exonerated, exhumed and interred in the floor of a cathedral in May of the year 2010.

But the professors of astronomy were not so forgiving! The church was merely tut-tutting the sad mistake of one of their own who had made an effort to think for himself. As for the astronomers, consider their plight. As youths they had elected to dedicate themselves to this most noble of all professions. And please recall that this was at the tag end of an era when many universities taught advanced classes in long division! If our present world had persisted in using the Roman system of numerals long division would still be an advanced university course, and if anyone doubts me I invite them to divide MMCDLVIII by XLVI without cheating by thinking in terms of the Arabic system (which probably originated in India). Not all universities of the era were so backwards, but many were.

Even with the introduction of Arabic numerals there were ongoing troubles with advanced calculations until the Indian concept of zero was added to the mix and modern positional mathematics became possible.

Returning from my little digression, after learning long division a budding young astronomer's course of study was followed by roughly two decades of intense labor where he was required to learn techniques for calculating the cycles, the epicycles, the deferents, the retrogrades and all the other motions of the planets as they orbit the Earth.

The whole professional life of the old astronomers had been dedicated to the art of determining and predicting the positions of the Sun, the moon and the planets. Now they are gray-bearded professors, accustomed to being looked up to in admiration and awe by virtually every student for their erudition and comprehension of this, the queen of all the sciences.

Practically overnight and straight out of the blue, some outside amateur steps in and tells everyone they are hopelessly ignorant! Within a few years flocks of callow youths are enrolling in their courses already knowing more about how to calculate planetary orbits than their professors do.

The sheer humiliation of it! Their whole lives are shattered! Even if they set out to learn the Copernican system they would still be hopelessly behind many of their freshman students! To these hidebound conservatives Copernicus was talking a heresy far more corrupting than mere skepticism about the nature of the Trinity! He was teaching nihilism, seeking to subvert the entire science of astronomy and skewering the very heart of Christianity!

He was trampling on their collective egos! And this was a sin far worse than simple blasphemy! The very name, Copernicus, was as bile in their hearts and mouths. He had to be eliminated at all costs!

An interesting subchapter in all this is the fact that in a test of the Copernican system against the Ptolemaic calculations the Copernican approach proved a dismal failure!

The astronomer could predict where Mars would stand in relation to background stars on the same date a year or more hence with an error of less than a degree whereas the best the Copernican system could eke out would be off by around three times as much! Not until Kepler demonstrated the worth of elliptical figures in determining orbits would the Copernican flaw be overcome.

This problem arose as a consequence of a subtle philosophical error; one which permeated the thought processes of literati throughout Europe at the time. This was the scientific dogma, initially propounded by the ancient Greeks, that God only worked with perfection; and ellipses were regarded as being imperfect circles; ergo by postulating elliptical orbits for the Earth and other planets Copernicus was claiming that God is imperfect!

Nothing could be more subversive of Christian doctrine than to begin by insisting that God is flawed! Not until Kepler's work established the reality of elliptical orbits was the Copernican system relieved of the need to postulate "perfect" circular orbits for the Earth and other planets.

But this is only half the Copernicus story; and the simplest half at that. The really significant element is still not truly appreciated.

And just what is this so important factor?

It lies in the fact that there is no real difference between the two systems!

Copernicus' work is a simple transformation of coordinates. If anyone chooses to do so, and is willing to spend the time and energy, he can prepare a system where Pluto is the center of the universe and everything revolves around it! It would be horrendously complicated but it can be done; and the mathematics would be perfectly correct!

To this day any seaman whose radio and GPS receiver is on the blink but who still owns a copy of the annual ephemeris and the classic "Dull & Dull" (No kidding, 'Dull & Dull' is its name…and it lives up to it in every respect) "Elements of Nautical Navigation" is still using the Ptolemaic astronomy. It persists to this very day and remained the primary method of sailing at sea, and transoceanic flight before Loran, as recently as 1950! The reason why it is used? Nothing esoteric; it is simply the most convenient approach and requires fewer computations. Basically, you use whatever mathematical origin works out most conveniently for your purposes. If you are concerned with astronomical position finding on Earth use a geocentric (i.e. Ptolemaic) system. If you are concerned with the solar system use a Copernican, solar-centric system. Want to play with a galaxy and pretend you are aboard the Enterprise and moving at warp 12 (2,048 times the speed of light)? If so the optimum reference base will be the galactic core. For a family of galaxies you assign a convenient focus at the approximate center of the cluster and go from there. For the universe as a whole (assuming we ever decide where its center is) you use that as the focal point.

In short, the whole thing is a tempest in a teapot. You simply adopt whatever coordinate system is most convenient for your purposes. It is utterly devoid of any hint of theoretical, theocratic or philosophical significance.

So old Copernicus was entirely right after all when he maintained that for purposes of determining the positions of the planets his approach, employing the Sun as the origin, would be more convenient than the Ptolemaic system. It took churchmen and scientists to turn it into a philosophical centerpiece for deciding questions of heresy... either theological or secular. Despite this minor caveat, Copernicus and Galileo surely deserve our unqualified respect and admiration. Between them they broke the shackles of theocratic arrogance and scientific bureaucracy and freed the human mind to explore the universe rationally. Mind you, this is not to say that everyone exercises that freedom; but enough do to make a real difference in a world still largely determined to hang on to fossilized stone age dogmas... both theological and scientific

Galileo stands firmly between the old physics and astronomy and the newer ones. While not truly modern he may be regarded as the bridge between the old and the new physics and astronomies. He was a new breed... a hands-on scientist who is noted for his endless experiments. He had an assistant drop two objects of differing weights simultaneously from the Leaning Tower of Pisa (The Campanile, or 'bell tower') while he wisely stood not too directly below and saw for himself that they fell at the same speeds; in the process disproving the Aristotelian assertion that objects of differing weights fall at different speeds. He developed the laws of kinetic energy by constructing gently sloping ramps so he could roll marbles down them and thus evolved the $K=1/2MV^2$ equation, which stands to this day. He constructed a telescope which enabled him to see the craters on the lunar surface and the rings around Saturn. And most importantly for the modern world, he observed spots on the Sun.

It was this, along with his adoption of Copernicus' assertion that the Earth moved about the sun instead of the other way around, which got him into trouble with the Church. As one crustily conservative Cardinal put it after seeing sunspots through a telescope, "I would rather believe that a looking glass should lie than believe that God's heavenly light should suffer from measles!", an attitude not unknown to our modern wowser-type preachers and their true believers when it comes to new ideas.

But here again there was division within the Church. The Pope quietly favored Galileo while the ultra-conservative wing of the Church abominated him as a dangerous heretic who was endangering churchly domination of society... a philosophical conflict which persists to this day in all parts of the world; including the United States (where we have the uncomfortable habit of electing such intellectually challenged people to the U.S. Congress --- presumably because they believe themselves far more intelligent than mere mortals).

This conflict goes far back into antiquity and is firmly rooted in the myth of "the golden age". According to the myth the world has progressively decayed from an initial age of perfection... a "Golden Age"... where everyone lived in harmony, no one had to labor and there was nothing but idyllic happiness and contentment. In Christo/Judaeo/Islamic terms this was the age embodied in the Garden of Eden. In Greco-Roman terms the world was rather more heavily populated but still idyllic.

Then discontent began to disturb the people. They were bored and wished to tinker and compete. Simple foot races grew into organized competition, with increasingly unruly fans screaming from the sidelines. War was still unknown but individual violence took root along with nationhood and the dissention which always attends a conflict of goals.

But it was restricted and lasted in this form for many years. Later generations alluded to this as the "Silver Age" while further decay led to the "Bronze Age" and thence to the "Iron Age", which was regarded as the age where mankind stands today (from the time of Greece at least until the time of Galileo); a doctrine clashing head on with the opposing, more liberal doctrine, which insisted that the world was steadily improving, embodying a conflict of philosophy which persists to the present day.

Given this philosophy it is not surprising to find conservatives recoiling in horror at the progressive decay of order on Earth. Starting with Columbus and proceeding to Newton, thence to Copernicus and finally to Galileo; Pandora's Box was opened and their only hope was to employ the most savagely draconian measures in order to slam it shut. And the justly feared Inquisition was precisely the dreaded weapon conveniently at their command. They were, in large measure, the Christian forerunners of the modern Islamic Taliban; just as the jihad is a mirror reflection of the Christian crusades and the American wowsers!

Galileo made two mistakes; both of which Copernicus avoided because of the latter's understanding of internal operations in the Church. Galileo relied upon natural rationality of mankind and the aegis of the Pope to shield him from the Inquisition.

As for the natural rationality of humankind, this has been proven time and again to be a cruel jest. "I don't know what to think; tell me what to think!" This has been the plaint of most humans since the beginning of their history. And this sort invariably looks to the most asinine leaders because they feel an affinity for them. They regard themselves as being cut from the same cloth... and all too often they are correct in their gut feelings. With only rare exceptions stupidity first attracts then selects stupidity, just as arrogance attracts and selects arrogance.

The Roman Church of that era was organized much as it is today. To a considerable degree it resembles the United States. A bishop or cardinal is in charge of his individual diocese, which corresponds to one of our states, and generally has absolute authority over all matters of interest to the Church. Among other things, he appoints the local Inquisitor for his diocese; he can set the tone for the appointment of priests and other laics and thereby nudge his diocese either to the left or the right.

The system is roughly paralleled today in the several States of the Union and the counties and cities within the states. The President, whoever he may be, is only marginally unrestricted in his decisions and actions.

In the Church it was routine that the priesthood at all levels was a lifetime appointment and the Bishop of Pisa, for instance, was accorded a free hand so long as he confined himself to activities regarded as licit for his diocese. Until fairly recently an appointment to the bishopric was an appointment for life so it was not unheard of for occasional prelates to lapse into senility but continue in nominal command while his staff actually ran things and made all the decisions in his name.

In the case of Galileo, while the Pope admired him, the bishop in whose diocese Pisa laid was ultraconservative and had several times had his Inquisitor issue writs of obstat to forbid publication or communication of his ideas by him. Always a quiet Papal nudge had saved him from too much inconvenience. But then, in a fit of foolishness or arrogance, Galileo published a book on the solar system. It was a fairly innocuous text, but it was couched in the fashion of the dialectical method of the era so the book consisted of a formal disputation between him and a straw man entitled "Simplissimus" (Latin for Simpleton" or "Fool")! Now he was indeed treading on dangerous ground.

Galileo's opponents anticipated modern 'spin doctors' and promptly took a copy of the book to the Pope and convinced him that Galileo was mocking him before the whole world! That was all it took. Galileo was tried and convicted, but even now the Pope sheltered him as much as he could without overstepping his traditional limits. After a largely pro forma recantation of his heretical views he was confined to his villa and forbidden from publishing any more papers or books. Something like 400 years later the Church rather belatedly confessed that it had made a mistake and cleared him of all charges. This explains a lot. But it does not explain why I regard Galileo as possibly the most important single figure in ancient *and* modern science. What makes him of special importance is the fact that he was willing and able to tackle the clichés of antiquity head-on. Two thousand years of blind faith in the authority of Aristotle were at stake. It was obvious that "the heavier the weight the faster it fell." Aristotle said it so it was true. "Women lack the lower pair of ribs." Aristotle said it so it was true. Etc.

Galileo was not satisfied to let the matter be. Instead, he had the audacity to go and find out for himself. The result was a shattering blow to the naïve faith in the perfection of the great Aristotle's reasoning. After lo, these many years the reigning monarch of natural philosophy was dethroned!

Nor was he content to second the Copernican "wink-wink, nudge-nudge" fiction that the Earth stood still in the heavens while the Sun circled about it and that the solarcentric mathematics was merely a means of simplifying the computations. His discovery of sun-spots, the rings of Saturn and the four main moons of Jupiter were just the icing on the cake.

In brief, Galileo anticipated the sentiments proclaimed by Sportin' Life in Porgy and Bess: "It ain't necessarily so!"

Just as modern conservatives remain outraged at Darwin, at contraception, abortion, politics, etc., the conservatives of Galileo's were increasingly dismayed. They were becoming the laughing stock of the literate world! Only one thing could stop this agent of Satan; trial and conviction by the Inquisition. He should be burned at the stake as a warning to liberal thinkers everywhere! Once the Pope hesitated they behaved like thud and blunder conservatives everywhere and lumbered full steam ahead while damning the torpedoes!

Convicted of heresy and faced with the alternative of abjuring his mistaken ideas or spending the rest of his life locked in a dungeon, Galileo wisely opted for abjuring. He dutifully proclaimed his acceptance of their censure and denied that the Earth moved in the heaven before the empanelled priests who has just convicted him.

He was thereupon retired to his villa and its surrounding acreage where he spent the remainder of his life in quiet retirement. But the damage was beyond control. Galileo's heresy spread like wildfire, and lacking any rational means of combating it the conservative prelates reacted in the only way they knew how... by taking refuge in ridicule and invective. Which was an even greater error? In cursing and ridiculing Galileo they overlooked the fact that they were also advertising his findings and that anyone who chose could replicate them for himself! The result was a nearly complete humiliation of churchly authority over the findings of physics and science.

It is one thing to debate endlessly such arcane subjects as how many angels can dance on the head of a pin, (which actually occurred and had a definite relevance in its context) but it is something else to insist that Aristotle was correct when he stated that women have one less rib pair than men. As the much earlier philosopher/monk, Duns Scotus, pointed out warningly: "The Church may indeed be wrong. The massed authority of the Church may insist that the sun rises in the west and sets in the east. But if I venture out of doors and see it rising in the east and setting in the west I cannot hold my peace. I must declare the Church in error; and if the Church is wise it will confess its error and resolve henceforward to confine itself to matters of the spirit."

Summing, Galileo's greatest merit lies not in his numerous stunningly fruitful creation of the scientific method and his discoveries while using it, but in the fact that he confronted Churchly and civil arrogance and authoritarianism head on... And won!

II
POST- GALILEO, PRE-EINSTEIN

Following Galileo, European science promptly took off on an orgy of discovery and invention from which it is only now starting to emerge. For the most part this is the sort of thing we should expect of cultures at long last ridded of the shackles of mind and body control. Most individuals will promptly seek to find someone who will enslave them anew so they can never be accused of anything and always have someone else to blame for their woes. A smaller number will cast aside all restraint and happily start setting themselves up to become the next stench of tyrants and are busily recruiting their own legions of serfs and peons. A rather lesser number will burst forth in a flowering of creativity which will resonate through history as a golden age of art, of literature, of science, etc. That is the way of us humans so we would be wise to accept it and factor it into all our equations.

I mention this period between Galileo and Einstein mainly to point out a few amusing illustrations of the persistence of error even after the error has been recognized for centuries.

Skip to the 'modern' era. Anyone ever heard of "phlogiston" or a chap named Wegener? Therein lay tales well worth the telling.

Galileo may have slain the dragon of ancient infallibility but, though slain, the dragon was not fully convinced it was dead. Ancient philosophy decreed that there were certain substances which could not be divided into mixtures with other substances. These fell into four categories, which were called "elements". Earth, air, fire and water were the names assigned them. Mix earth and water together and you wind up with mud. Boil the water away and you've got your earth back again while the water can be recovered from the steam. The whole 'science' of alchemy was premised on this antique science. Mix mercury and lead, antimony with silver, or whatever, in the proper manner and you come up with gold, etc. The all too natural fondness and avarice of men encouraged them in this faith, but it soon ran into a problem. Local alchemists faced a conundrum when they sought to reduce gold into ordinary dirt, iron into copper or copper into lead. It gradually dawned on them that it couldn't be done. These were elements in their own right! So now we have gold, silver, copper, iron, earth, air, fire and water as elements. And that didn't make sense.

More experiments and before long elements were being strewn all over the landscape. Everywhere one looked there was some alchemist stumbling across a new element. Talk about sharks enjoying their little feeding frenzies or priests rooting out heretics or lawyers chasing ambulances, the alchemists showed sharks and priests and lawyers a thing or two about frenzies in their frantic scratching after new elements.

So far as I know there were no Galileos lurking in the wings proclaiming anew the demise of Aristotle and a new era of science. There seems to have been a sort of quiet consensus beginning to appear more or less spontaneously; that the ancients may have been wrong here too. The first tentative steps favored Aristotle sort of focused on the problems with water.

Dissolve salt in water and you get salt water. But boil the water away and you have your salt back unchanged. Not too promising. Add to this the fact that the water which was boiled away comes back to you as pure water once you set up the necessary collection procedure. So perhaps water actually was an element. But then came a distinct hitch in those efforts.

Mix a pint of alcohol with a pint of water and measure the volume. You don't get two pints! It is more like one and four-fifths pints! And adding to the problem, the weight of the mixture exactly equals the weight of a pint of water plus a pint of alcohol when measured separately! As might be expected, it was the distillers in Scotland who first spotted this peculiarity; or at least, so I have been told --- which make perfect sense when you stop to think of it. Who better to note the problem than those kindly gentlemen who worked with alcohol and water in massive quantities?

But it still defied rationality. Add one and one and you get two; not one and eight tenths! But alcohol plus water clearly violates the rules of mathematics here; and mathematics were sacrosanct... as any kabbalist or modern physicist will cheerfully attest to all and sundry of any audience.

Adding to the problem, if you mix a little electrical charge to a pan of water the water slowly disappears, presumably having been converted into air. And worst of all, the converted air turns out to be highly flammable, and by burning it you recover approximately the same volume of water you started with!

This did not make a great deal of sense.

It would nice to believe I am only imagining the conundrum posed by the behavior of water to the slowly growing body of chemists of the period, but this would be unrealistic. Alchemy was an accepted science until roughly around 1600.

Chemistry only started to become a true science in the late 1700's. Until then there was a highly blurred distinction between it and alchemistry (sic). These chemist/alchemists did not leave copious notes, especially of their unresolved quandaries or odd little glitches in their theories. Thy did not wish to risk forfeiting the fame and fortune which would accrue to the successful alchemist who could solve a puzzle and perhaps make a breakthrough on a problem; and publishing notes of their failed experiments might well steer some rival onto the right track. So I am engaging in sheer speculation when I recount the likely progress of the science during this era. But I cannot believe I am far off base. It is the way such things work; it is the way of the world: Its tao, if you will.

The discovery that low voltage direct current electricity splits water into two different components, both of which were gasses, was the last straw. Earth, air and water had been proved as mere compounds rather than elements.

But what about fire? Something was very wrong here. The supply of fire appeared to be inexhaustible. You did something with a flint and steel. Sparks of fire emerged and you could light your kindling with it. But neither the flint nor the steel were much changed; and the next time you used it there was a fresh fire.

Where did this fire come from, and why could you do nothing with it apart from creating new fire? Where did it go when it left? Clearly it stood apart from the other three elements. How and why? Those were the questions.

I suppose there was occasional experimentation by scientists hoping to gain fame and fortune by unlocking the secrets of fire. But most of this merely resulted in burned fingers... if not worse. Ultimately science hit upon the idea that fire was the exception to the rule: that it was truly an element. And being an element meant it ought to have a fresh name; a name which would distinguish it from the plebian 'fire' and contribute to the arcane mystique of their occult science. "Phlogiston" was the name selected for this elemental fire. In due course entire avalanches of paper devoted to mathematical analyses of this magic stuff, all of them purporting to show in detail how it acted and reacted in the outside world!

The interesting point of this is the fact that, as a rule the mathematics was all quite correct. They were correct even though there was no such thing as phlogiston, nor had any of it any relation to the mundane phenomenon of fire!

The scientists had simply fallen into a trap of their own making. They had proved they were humans by coming to worship at the altar of their own creation. To explain this properly we must allude to one of the core beliefs of many otherwise pretty good scientists; i.e. that the universe is so vast and so varied that every possible alternative must be real – at least somewhere and some time! Given this basis, phlogiston must be real, providing we can give it mathematical support which applies against every observed condition of fire.

Such support was soon produced via a blizzard of mathematical papers purporting to define the needed characteristics of phlogiston and its hypothetical connection with the universe. Once this was spelled out in detail phlogiston was regarded as proved, and thousands of scientists the world over were utterly convinced that the beauty of mathematical analysis had once more unlocked one of the innermost secrets of the universe.

Once again, science was at long last verging on a complete solution to the mystery of everything! All that remained were a few odds and ends to wrap up the universe once and for all. Then we could forget about it.

Of course this was not true. What was overlooked then and continues in large part to be ignored today is fact that many exotic 'truths' can be discovered if we begin by instructing our computers to spell out the relationships which must exist; providing we spell out the factors and require the computers to construct mathematical bridges across all inconvenient gaps between observed bits of actual data.

It is the G.I.G.O. statement all over again; and every computer user knows it. You put garbage in and you get garbage out! And this applies to pen and pencil calculations as well as computers.

There is an element of lineal thinking here. To illustrate by simplifying to the point of absurdity, we can arrive at 100 by laying out 100 consecutive 1's, or 50 2's, or 25 4's. Or we can make it far more impressive by writing $100 = 2^2 \times 2^4 - 2^3 \times -1^2 = x^2 = 100$. If we really want to impress the yokels with our brilliance we can reduce the whole equation to chaos theory and come up with a 60-page monograph.

But they all amount to the same thing. In general, the greater the complexity and inflation of the mathematics the more likely it is that the culprit is indulging himself in a bit of intellectual bullying redolent of Arnold Toynbee in his "Study of History", where he snobbishly litters his work with extended quotes in Latin, Greek and Arabic, *et al* while studiously avoiding translations.

This is precisely the tactic employed by alchemists, priesthoods, P.T. Barnum's, used car salesmen, politicians, lawyers, astrologers and moss-bound academics the world over. Make it sound arcane enough and you can rope in the suckers on a wholesale basis. To cover their tracks the hucksters take refuge by proclaiming their own divine omniscience and freedom from the prospect of error.

There is an old saying, "If it works, its right. You can't argue with success!" Sorry about that; you can argue with success if it is merely an over-hyped equation. There is even an unhealthy element of wishful thinking admixed in our mathematical interpretation of mathematics; as when the identical detailed observations are shown mathematically to arrive at diametrically opposite conclusions!

A current example of this may be seen in the patterns of fractures on the lunar surface. In the 1930s P.A.M. Dirac used these rifts to argue that the gravitational constant had decreased over the aeons so the lunar surface cracked open from the expansion of its circumference... rather like stretch marks on a woman's stomach just before delivering a child. But by the year 2010 these same fractures were being attributed to shrinkage of the lunar circumference caused by folding! --- Same data, opposite conclusions; and both supported by the same mathematics!

Amusingly, there is a strong likelihood neither is correct and the rifts are lava tubes relict of the thermal activity and possibly caused by heat generated consequent to major meteor impacts. Yet a fourth alternative would have it that the cracks are the consequence of tidal friction created by the Earth's libration. But no matter, the point is clear. Mathematical equations may prove to be a two-edged sword if we worship them too much.

On the flip side of this is the almost obsessive determination to reject any idea for which the mathematics is not perfected. A recent example of this showed up in the 1920's when a geologist by the name of Wegener pointed out that the coasts of Europe and Africa were almost perfectly symmetrical with the coasts of North and South America, a match which was made even more pronounced if the continental demarcation lines were drawn along the continental shelf – where relatively shallow water abruptly plunges thousands of feet to the true floor of the ocean.

It was further reinforced by the observation that certain land features, such as bluffs overlooking the ocean on both continents, were symmetrical and exhibited the same stratificationing.

Wegener proposed that the two were initially a single continent which split and drifted apart a few aeons ago.

Lacking a completed mathematical model (which could not have been developed with the then-existing data) Wegener was solidly mocked and hooted down by the physics fraternity. His reputation was demolished by place-proud colleagues who commiserated pityingly over the 'tragic' fate of an otherwise promising scientist who turned out to be a crackpot! Luckily, Wegener lived to see himself thoroughly vindicated and today the concept of continental drift is securely anchored in the best physics texts; which goes to show that an array of mathematics and the fanatic backing of over-zealous practitioners of science can make the worst tasting of medicines go down. Truly, churchmen prefer to employ the stake to dispose of heretics, but the scientist is crueler; he does it with ridicule and a sneer.

Another modern day version of this, but with a twist, is the tachyon. It seems a physicist was playing around with some mathematics relating to the speed of light and in the process he realized that the completed equation required the existence of particles which flitted about faster than light!

When a missile, or whatever, in our universe goes from some presumed speed and accelerates to the speed of light the other side of the equation must consist of supraluminal particles which begin with the speed of light and allows these supraluminal tachyons to accelerate from there, presumably to a new speed of light on the other side of ours; or, arguably in a deceleration back to zero on the right hand side of the equation!! One might legitimately head off in the other direction and argue that we are the tachyons and below is a spectrum of "snailyons" whose top speed is the speed of light and whose slowest speed is some sort of zero down below! ....Except that it doesn't work. It fails to consider the question of 0^0 Kelvin where all molecular motion ceases. Nor does it contemplate the possibility that when these tachyons commence decelerating them merely falls back unto this universe, regain their speed here and continue repeating this oscillation eternally. In short, from any rational viewpoint the tachyon model seems terminally flawed.

"From the presence of a watch I infer the existence of a watchmaker". This little tidbit dates back to the late middle ages but continues to becloud the perspective of our physics boys to the present day.

It is perhaps best exemplified in Wagner's entry procession of the god's into Valhalla in his Ring cycle. The grandeur of the music is solemn and awe inspiring. A glorious inevitability is implied; an inevitability which was expressed by the churchly rejection of Copernicus because God only deals with perfection and elliptical orbits are imperfect circles!

In modern terms we watch it evolve into the recent obsessive effort to discover a direct lineage of man as presented in Disney's "Fantasia", which depicts a fish emerging from the ocean to become an amphibian, then morphing into a dinosaur, thence to a gorilla and from that into a man.

Reality points in the opposite direction. Contrary to human prejudices, which seek to present a vision of the universe which is majestic and perfectly organized, in actuality we exist in a messy and disorderly universe. There is nothing simple and straightforward about it, and if there is a God it must have a delicious sense of humor or it would never have created the duck-billed platypus or permitted theologians to provoke mankind in their effort to acquire authority.

I would be less than forthcoming if I failed to cite some of the more unsavory elements in the science equations. Politics plays a major part in its version of science... witness the various Protestant theological outrage over claims that the Earth is entering a phase of humanly manufactured global warming or the arguments against evolution or the 'sin' of stem-cell research... and the Soviet Union's espousal of Lysenko and his little theory of acquired evolution. As an added for instance; the whole period between Galileo and Einstein is jam-packed with discovery in every direction and discipline. It boasted its share of flawed theories created by inadequate knowledge. But this was only to be expected of a wide open expanse of a newly discovered universe of knowledge never even hinted at in the works of earlier generations. Everything was in a flux and nothing seemed permanent, save for a few new laws of nature, such as Newton's trio, etc.

Something new had come into the world and everyone wanted to get in on the action. Which mainly goes to show that mathematicians and scientists are just ordinary humans like all the rest of us? But unhappily, this 'all the rest of us' includes preachers who somehow contrive to know with utter certainty the will of their God and seek to impose their prejudices on the society, as well as the maunderings of Congressmen and kindred vulgarians who are convinced that not only are they qualified as lawmakers, but equally so as economists, generals, admirals, physicians, engineers, and in every branch of art and science!

By now I **expect** the bulk of my audience has given up in disgust at my preachments and failure to address the state of the universe as well as my refusal to adhere to the arcana of conventional scientific jargon. This was deliberate on my part and meant to lay the groundwork for criticism of the failures of pedagogical presentation.

As far back as the seventh century b.c.e. the Greek philosopher Hesiod pointed out that in when it came to learning, the process of learning was more important than the subject to be learned.

M.I. Finley probably expressed this best in his "The Ancient Greeks" Peregrine Books, 1963, pp

114). Speaking of the Greeks of Plato's age he cited Hesiod, saying as an anticipitory echo; "The servant had become the master: the manner in which an idea was expressed had become more important than the idea itself. This was already too evident in the fourth century; it became the rule in the Hellenic age."

It remains so to this day. Sadly, after more than 2,700 years we have come a full circle where the form in which an idea is presented is more important than the idea itself. So, in what some may regard as insolent disdain, I have elected to throw over the traces and focus on concepts and the logic underlying them, hence my seeming disjointedness.

After all, every generation of scientists labor under the convenient delusion that they knows the *Truth* and it only needs a bit of rounding off and touching up before it is revealed in its full glory! And every generation has fallen flat on its face before getting up, brushing itself off and shouting loudly so all can hear, "Now we're finally on the right track! It is certain we'll get it right tomorrow!" Is it too much to expect a touch of humility from the science community even if they have to force the pretense?

This completes my apologia. Now it's down to business.

.

III
ZERO, Etc.

"Mommy" asks the little cave boy, "where did I come from?"

"You came from mommy's tummy."

"How did I get into your tummy?"

"A God came by and saw mommy busy playing with the other children. The god thought to himself, 'This girl is old enough to begin working so she can keep me and all my priests fed and help save her people. 'So he planted a little seed in mommy's tummy to make you grow and come out of her. Now mommy has to work all the time and never gets any rest."

"Where did the god come from mommy? What does a god look like? Why can't I see one? I want to see a god! How did he plant the seed in mommy's tummy?"

"He was always here. Now stop asking foolish questions and go out and play with the other children, who don't go around pestering their mommies with endless questions."

A few thousand years later the questions were more pointed and were being asked by older thinkers.

In India the world was perched atop the shell of an enormous turtle which swam in the endless ocean.

What was on the other side of this endless ocean?   Don't ask.

In ancient Greece the first gods were the Titans. The eldest titan, Kronos, begat Zeus who in turn begat yet other gods to create the pantheon.

Who created the Titans?

Ask all you want; we're not going to give any answer that makes sense. Modern religions are much the same. Call It Ptah, Yahweh, Yahoo, Jehovah, Allah, P.T. Barnum or the Great Panjandrum; by whatever name we use, it created the universe? Who created this god?

Nobody. He just was.

Alternatively, some have argued that this God simply willed himself into existence!

Huh!? Run that one by me again. First there was nothing, then this nothing decided it wanted to be something so it manufactured itself!

Makes a lot of 'cents', I suppose... enough that untold billions of people have donated hundreds of billions of cents supporting the people who preach it! Some cents; some sense!   In ancient Sumerian times the Land of Dilmun, which today is called Bahrain, was called the "Land of the two oceans" because, in the midst of the salt sea it bubbles up streams of fresh water. To the philosophers of the era this was regarded as proof that the salty ocean floated atop the Earth, which floated atop an ocean of fresh water.

Actually this was a highly sophisticated scientific explanation, given the data available to humankind some 5,000 years ago. It is doubtful many modern physics practitioners could do as well were they transported back to that era and were left with only the knowledge available at the time.

In ancient Egypt Ptah was the god who cast dirt into the ocean until it rose above the sea and became an enormous mountain. He was pretty much the father of the Egyptian gods and retained a considerable importance in the Egyptian religion even though he was supplanted by Amon/Ra and Osiris/Apis in popularity and authority. Always there was this emphasis in pushing back the limits... but never was there any real success.   For   entire millennia fresh ideas were singularly lacking.   I suspect the human mind simply is not capable of contemplating infinity. It recoils upon itself and takes refuge in platitudes.

By the time of the Egyptians a number of tangible infinities had been placed on the table. There was the infinity of space. Visualize the universe around us. Do the stars go on forever and ever? Say yes and now we are obligated to ask ourselves what is on the other side of 'forever'?

Without limits; just going on forever and ever. We find ourselves having to say there is nothingness outside, but then we find ourselves asking where the nothingness came from!? Or Einstein, who proclaimed that the universe is "finite but unbounded"? It sounds impressive but it takes us precisely nowhere.

How about time? Ask just anyone. There is an eternity of time waiting to happen plus another eternity of time which has already happened. And here again the Greeks crop up on center stage. The true creator Titan, the one who began the whole affair, was named 'Kronos'. And what does Kronos translate as?

Would you be willing to hazard a guess that Kronos translates as the Greek word for 'Time?' If so, give yourself a cigar. This is the reason we now enjoy the disciplines of 'chronology' for dating events; with which to plague young children in their classrooms by compelling them to remember such arcane data as "Fulton tried out his steamboat on August 11, 1807 (if I recall correctly from my school days.)

Then there is length, width and height; all of which can be extended to infinity.

It gets even more interesting when we delve into the intricacies of modern mathematics and physics.

I mention these little snippets of ancient thought processes and conversations to illustrate the antiquity of the problem. It has hung around since the time of the cave man: and always without resolution. To repeat myself, the human mind simply is not really capable of comprehending infinity. We spend too much of our time seeking to trivialize reality… to bring it down to our level… i.e. roughly on a par with the latest rock band or pop singer. But infinity refuses to be trivialized. How many times do we hear some media hack proclaim that the popular actor, musician, rock star, or whatever will be remembered "forever"? Generally such types are lucky if they are remembered for more than a century. And most business tycoons or congressmen are forgotten in little more than a decade.

And as for the latest billionaire who made his pile by mulcting suckers via the stock market or inflating bank charges ---that sort of greedy parasitic hog is forgotten so quickly it is pathetic! On the positive side, over the years we have been able to refine our definitions of infinities and have learned how to manipulate them to a degree.

For example, take the infinity of numbers. Start with a 1, then add a zero to make ten. Add another zero and you get 100, then 1,000, 10,000, 100,000 etc. No matter how many zeros you add you can always add another zero. And when you start to run out of paper you simply escalate by squaring it, then escalate anew by squaring the

40

square. No matter what we do there is always more that can be done. Psychologically, we cannot even conceive of a limit to the process.

But this does not mean we have said everything of interest or utility anent the infinity of numbers. We can subdivide it(?) by extracting only even numbers, 2, 4, 6, 8, etc. So we have infinity of even numbers. We also have infinities of odd numbers, of numbers divisible by 3, or 5, or 7, or whatever.

Whoa now!

Intuitively, it is obvious that an infinity of numbers divisible by 5 can be only 1/5 the size of the infinity of 1's. Doesn't this mean that the infinity of 1's is five times larger than the infinity of 5's? Not at all. Let's run a little chart to prove my statement:

```
1,  2, 3,  4, 5,  6, 7,  8, 9, 10 (11) etc.
2,  4, 6,  8, 10, 12,14,16,18, 20 (22) etc.
3,  6, 9,  12, 15,18, 21, 24,27,30 (33) etc.
4,  8, 12,16, 20, 24, 28,32,36. 40 (44) etc
5, 10, 15,20,25,30,35,40,45,50 (55) etc.
```

Clearly, for every digit on the 1,2,3, scale we can add a digit on the 2,4,6, the 3,6,9, the 4, 8, 12 and the 5,10,15, scales. We can also see where the same must hold true no matter where we pick up on the 1,2,3, infinity progression.

In short, every permutation derivable from any of these, or any other progressions are merely parts of a single infinity, i.e., the infinity of numbers.

Carry this step further. The class of all negative numbers as well as the class of all fractions may be included in the same infinity, e.g., the class of all numbers, *period!*

What have we done? We have thrown a hurricane of numbers of any sort into a single basket and declared it to be an all-embracing infinity. But we have also done something else without realizing it. Casual manipulations of infinities do not alter those infinities. For example, from our little chart try to divide any infinity by 5 or 10 or 1,000. This is merely the reverse of multiplying, which is simply another form of addition. Even dividing an infinity by infinity leaves us with infinity! The effect of this is to render all ordinary manipulations of infinity of null effect. Infinity squared is infinity. The square root of infinity is infinity, etc.

If all this seems a little arcane…it is! And I apologize for it.

The trouble is, the concept of infinity is so central to many of the problems afflicting cosmology that at least a marginal acquaintance with the logic of the genre is essential if you are to comprehend the universe as a whole.

Getting practical, we can hack into any infinity and perform a series of calculations which are meaningful. To illustrate, we can designate a 1, a 2, a 3, etc., and construct a magnificently profound edifice of logic from it. The segment we have selected has no relationship to the underlying infinity save for the fact it is a tiny sliver of the whole, but we can create a meaningful zip code for our postage without concerning ourselves that we may have actually toyed with infinity. We can create geometries without claiming that the areas measured have some deep significance to infinity. We can give everyone a social security number and not affect infinity. For that matter, we could assign a different 10,000-digit identity code to every atom in the universe and still have infinity of numbers left. Infinity simply is.

In 1895 a German mathematician named Georg Cantor published the initial part of his "Contributions to the Founding of the Theory of Transfinite Numbers." In 1897 he completed his text. It is a slender book of some 210 pages in the paperbound Dover publication dating back to 1915. I am reasonably certain there have been further editions since then but have not bothered to track them down since there would be no point to it. The work is frequently cited in advanced textbooks of mathematics but it appears that Herr Cantor not only contributed to the founding of the discipline; he also completed it. At any rate I know of no follow-up texts of any consequence which have been published since then, so my dog-eared copy is probably pretty much a collector's item; providing anyone is into collecting such arcane objects.

One interesting product of Cantor's work was not directly addressed in it but is clearly extractable from it. It concerns the interaction between the applications of infinity to a finite situation. Examples include the Euclidian failure to square the circle, and the problem of the relation of a side to the diagonal of a square. These are geometrical as well as mathematical problems so we are returning to realities in a way.

Imagine that! You have my permission to be astonished. At long last I am getting around to the relevance of my initial reference to Pythagoras which opened this chapter!

The Pythagorean Academy (which obviously was named after its presumptive founder) removed itself from Greece to the boot of Italy, again presumptively to insulate their members from the frivolous, highly politicized hurley-burley of the Greek heartland and free them to enjoy the bucolic pleasures of a contemplative life... not to mention the more erotically mundane pleasures of sybaritic life... which just chanced to be named after the city of Sybaris... where the Pythagoreans soon learned to adjust to the local customs and appreciate both the erotic and technical nuances of their orgies.

What little we know of the Academy is of dubious value, but well worth passing on, if for no other reason than its insight into the character of its founder's digestive system (or perhaps in deference to the olfactory sophistication of the native Sybarites). Admission to the society was strictly contingent upon an oath to eat no beans!

We also know that certain numbers were closely guarded secrets which they dared not divulge to the uninitiated lest the word get out and cause the world to implode... or perhaps deform into something even worse.

Today we know them as 'irrational' numbers because they have no ends. When you try to calculate them exactly they defy all efforts and just keep trudging on forever with no sense of progression or long repeating series such as...252525, or 33333.

I believe pi has now been carried to 600,000 or so decimal places, with no hint of an end or repetitive series. It may be even more but there is scant point in trying to keep track after the first few thousand.

To the intensely rational Greeks, the idea of a bunch of irrational numbers darting nakedly across a page of sober calculations was akin to blasphemy. Yet there before their very eyes were two of these numbers perched squarely in the midst of sedately prim and proper numbers was horrifying beyond measure! Two renegade numbers are Π and the square root of 2 =1.414214+. If you have never concerned yourself with such fundamental arcana you might keep these in mind.

There is an infinity of related irrational numbers available, i.e., pi - $\sqrt{2}$, pi x 2 pi x -2, pi2$^+$ and a myriad variations, but these two are to my immediate knowledge the only ones of interest here. To the Greeks they were ominous; numbers which seemingly continued forever and ever without any hint of logic attached to them... which, in point of fact, is exactly what they do. But no matter which way you try to interpret the preceding paragraph, it is only partly correct. The numbers may go on forever but they are not illogical; irrational, yes, but illogical, no. Logic is there, but it is surprisingly subtle and, if applied properly, resolves a knot of thorny little problems which have pestered mankind for centuries. "Squaring the circle" is one such problem. There may be others but this is pretty central today.

Cantor began by defining any set of infinite integers as a "c set", or continuum. Examples might include the set of all numbers divisible by 2, or another set consisting of all numbers resulting from the multiplication of 79. If that continuum cannot be subsumed by any other continuum it becomes, in effect, an independent infinity. Now forget the fancy language. To a mathematician "continuum" is essentially another word for "infinity". So call it a "c set" and go from there. This is where the 'c' came from in Einstein's little equation e=mc^2

Now we apply a touch of Cantorian logic. Some infinities are of differing continuums. The set of all straight lines, for instance, is demonstrably different from the set of all curves since the two can contact each other only as discrete, and dimensionless, points. But I will return to this later. First we clear the deck by introducing a few more infinities.

Cantor's interest was focused primarily on mathematical infinities. But this does not exclude physical infinities which may exist. How about an infinity of speed… which is Einstein's "c" [which I will get to later]? Infinity of time? Infinity of space? Infinity of mass? Infinity of temperature? Plus a few other possibilities which may sign the log, i.e. an infinite array of straight lines, an infinite array of curves, an infinity of different shapes. There are no doubt others, but these will do for starters.

For the nonce, the infinities of lines and curves are of particular interest. Begin by visualizing a solitary point in space. From this point we draw an "infinite" number of radial lines extending in every direction. These lines are defined as having (what else?) an infinitely narrow width so there is no limit to the number which can be manufactured radiating out from the point. For curves we start with the simplest of all curves; a circle. A circle has neither a beginning nor an end and it goes on forever; it is therefore a curved infinity of the simplest possible variety.

Now draw a diameter to bisect the circle. To obtain the circumference of this circle we multiply the diameter by pi, and in the process try to convert a finite line into an infinite set of numerals. It therefore becomes an infinite subset of the infinity of line. It is designated by the Greek "pi". Note that this transformation must exist whether or not the diagonal is ever constructed. It is implicit in the act of drawing a circle. In effect, every 'o' in this manuscript is infinite, and no effort to deny this truth can succeed. This is all perfectly normal, so suppose we play a really nasty trick on those who seek to reduce the rest of us to cowering little ants by the acumen of their knowledge and sublimity of their calculations. Draw a circle and inscribe a diameter inside. From the junctures of this diameter with the perimeter of the circle throw out a pair of 45° lines to complete a right triangle where the diameter of the circle becomes the hypotenuse of the triangle.

Bingo! If *pi* were a rational number then a curve would be merely an offshoot of the set of straight lines, which is a contradiction in terms. Inasmuch as the hypotenuse of a right triangle is also the diameter of a circle, then if it were a rational number the diameter of the circle necessarily becomes a rational number.

The length of the hypotenuse is: the diameter of our circle times the square root of 2, or 1.414214+. But the operational number is D times 3.1415927+! And since the right triangle is inscribed within the circle as its diameter it is clear that squaring the circle is impossible. Two distinct infinities are opposing one another.

And it does not require a 40 page long equation to prove this conclusion... merely a touch of logic!

Lest some intemperate soul seek to argue that the diameter of a circle does not have to be infinite since we know it is the circle which is infinite we can point out that it is necessarily impossible to achieve an accurate finite measurement of an infinite body, so our initial conclusion is not invalidated. We may further generalize this statement by pointing out that any attempt to generate a totally accurate relationship between any chord on a circle or curve is destined to fail for the same reason. We will be seeking to compare one c set in terms of a different one, and two infinities cannot be resident of the same c set. As an even more inclusive generalization, this reasoning may be expanded to cover every conceivable c set, e.g. Zeno's paradoxes, et al. By definition, they are incommensurable with one another.

If this seems like an inordinate amount of energy to devote to a trivial subject, to a certain degree it is. But there is method to my madness, so bear with me a bit longer. Of sheer necessity any realistic cosmology must deal frequently with various infinities'. Like a dog gnawing over a bone, Cosmology itself gnaws away first at one infinity and then turns to snap at another one. Failure to deal explicitly with this truth has been a besetting failing of cosmetologists (pun intended). Too often it compels them to apply mascara to their clumsy cosmologies in order to conceal the issue; to plaster over problems rather than solving them.

Four final little items ought to be mentioned before I head into the main body of my dissertation. Since they fall mostly within the intermediate period between Galileo and Einstein this seems an appropriate time to mention them, even though their roots date back a thousand years and the fruits are only now starting to become a realistic factor in our thinking. Accordingly, I start at the beginning, continue to the end, and hope I have enough sense to stop once I have reached that point. The first story is interesting in its own right. No one really knows when it started, nor do we have any proof of who started it, though some fairly strong suspicions place it in India, possibly during the Asoka Empire. We know it crept into Europe during the Crusades, along with 80 to 90 percent of what today we still think of as our culture.

Of course I am alluding to Arabic numerals whose specially designated symbols which represent the numbers 1,2,3,4, 5, 6, 7, 8, and 9; with 0 to be introduced somewhat later. Had the Arabic numerals, or their equivalent, not been introduced the modern world could not have come into existence! I say this flatly and without qualification. Look back to the year 999 in Europe and reintroduce its science here today. That is approximately what we would be looking at were it not for the generosity of the Arabs; and probably before them, of the Indians of India. The Black Death would be a recurrent event, witches would still be burned at the stake alongside heretics. People would still be thinking about "tooth worms" which caused cavities, etc.

Today we regard the zero as primarily a means of establishing and maintaining positional relationships. Thus we have 10 being clearly distinguished from 100 and 100 from 1,000. In the old Roman system the corresponding numbers would be written 'x', 'c', 'm', etc., with 'I', 'V', 'L' and 'D' thrown in when the inconvenience of writing cccccccc rather than 800 proved unacceptable.

But this is only part of the story. With '0' now being a number it became possible to speak of minus numbers... a fact which anguished many of the more hidebound mathematicians no end. "I have one dollar in my pocket," was a common growl of the radical conservatives. "How could I possibly have minus one dollar in my pocket? That is not a rational statement." Or perhaps they were more general. "I am walking along at 1 mile per hour.

There is no way to argue that I am walking along at 0 miles per hour or minus one mile per hour."

Echoes of this argument were still hanging around during the period of the American Revolution, and I suspect there are probably a few determined protestors still lurking in the halls of Congress as well as remoter shadows of Appalachia or Mississippi or the Dakotas, among the majority of American congressmen or some of the more obscure fundamentalist churches.

But '0' is itself an infinity. It starts nowhere and ends nowhere, and whenever someone seeks to add or subtract something from it they come up short. Which proves it to be a member of a unique infinity in its own right? In other words, it is the parent of all the curves in the Cantorian C set.

I'm throwing in the following little item primarily for the fun of it, but also because in no small part, it illustrates the mechanics of a too literal mind set which runs along narrow gauge railroad tracks and is instantly derailed when something unexpected comes along, viz, How many angels can dance on the head of a pin?

This was actually discussed among the philosophical (which is to say, the scientific/Churchly) profession during the early and middle centuries of the Christian era, when the first universities were beginning to coalesce. At the time only males were admitted to the universities, and in order to matriculate they first had to take minor orders in the Church. I don't know what these orders were, nor do I much care. They may have been ordained deacons or presbyters or whatever, but in the end it resulted in them being called 'clerics', and they were highly prized individuals since they could both read and write. In due course they came to permeate society at all levels and the name 'cleric', or 'clerk' has become enshrined in the mercantile world. It was even held in such esteem that people wound up naming their children after the family profession, with the name 'Clark' becoming commonplace throughout Europe.

Be this as it may, the universities initially offered only two tracks toward a degree. The 'trivium' consisted of 'revealed' theology, philosophy, and either law or rhetoric; or perhaps something else along these lines. It has been more than sixty years since my attentions briefly touched upon the subject and I do not regard it as important enough to justify tracking down the exact order of study merely to boast of my erudition and smugly intimidate the unwashed masses along with the illiterate nobility who required my services to count their treasures and treat their gout.

The quadrivium was more complex by virtue of including what small amount of science was actually known at the time --- including, as I already remarked, special classes in long division. The quadrivium was accorded a degree of respect by the ordinary yeomanry, but the esoterica of the churchmen, with their incessant debates over the minuti of divine activity converted the trivium into a mockery called 'trivial' as in 'it is a trivial argument' as the modern echo of these studies... and to a great degree it is a well-deserved echo. But it was not always entirely deserved, and the 'how many angels' debate is a case in point. And here again the roots lay about two thousand years earlier with the Pythagoreans... or perhaps with Zeno and his paradoxes... both of which pestered philosophers and mathematicians until as recently as the 1930's.

Consider the multitude of angels first. Any infinity can be broken into innumerable parts, with each being infinite in its own right while also part of the c-set of all infinitudes of the same ilk. An inch long line can be broken into precisely the same number of discrete intervals as a line stretching across the universe (always providing the lines are postulated as consisting of dimensionless points).

Next consider the fact that God appeared to Moses in the burning bush in the Pentateuch (Remember, these were the middle ages when even the most minor paradoxes were ignored as the product of divine caprice). But this did not divert him from his meddling with events all around his constricted little universe, ergo, a woman in India is dying in childbirth, a Nubian in distant Ethiope is being fanged by an asp, a heretic in Spain is being burned at the stake while in a nearby dungeon a sadistic priest is superintending the cranking the wheel on a suspected heretic in Germany and a Philistine in the Levant is purchasing a new concubine.

We may even stretch this a little further and point out that divinity also coexists in all times so at the same instant this God is also supervising a terrorist blowing up the World Trade Center, or a rocket aimed at Pluto, along with a petty thief filching alms from a church box or a Cro-Magnon man bashing the skull of a Neanderthal woman as a sign of affection. This God is even supervising a fish as it crawls out of the ocean and learns how to live in dry land or an amoeba learning how to partner with a second amoeba to create the first multicellular life form. God is even superintending the final gasp of life on Earth as the sun goes nova! These and a myriad other details were being attended to at the same instant, and in perfect harmony to the divine intent. Truly, God was at one end of a most remarkable party line in the communications industry. Given these conditions it was inevitable that infinity be invoked and intensely scrutinized and adjusted for logical consistency.

Given also that these were the Dark Ages it was equally inevitable that the results would be couched in theological jargon. Angels were entities resident of the empyrean and thus were created as infinite beings. Accordingly, they were so enormous that the whole Earth was as a grain of sand beneath their feet, but simultaneously so small that the whole host of heaven could dance upon the head of a pin!

Problem solved! Angels were never the focus of the question. They were merely there to exemplify the character of infinity. The subject was not at all trivial but outside scoffers and sneerers seized upon it like modern politicians and people with 'causes' of their own and distorted it to their own ends; which gave it the appearance of being trivial.

Zeno's little paradoxes also illustrate yet another aspect of transfinite mathematics. Briefly, and in paraphrase to fit my purpose: Achilles sets out to race a tortoise. But because the tortoise is so slow it is agreed that the tortoise will start at a point precisely halfway to the goal line. They both set off at the same instant. By coincidence Achilles arrives at the tortoise's starting point at the same instant the tortoise reaches the midpoint of his distance to the finish line. A brief time later Achilles reaches the point the tortoise was at when it was midway to the goal and finds that the tortoise is now midway ahead of him en route the goal, etc. Ultimately, Achilles is a minuscule fraction of an inch from the goal, but the tortoise is still half the distance to the goal ahead of him. No matter how long the race continues Achilles can never overtake the tortoise....but the tortoise can never reach the goal either!.

A variation on this has Achilles shooting an arrow at a target. It covers one half the distance in 1 second, half the remaining distance in half a second, half of the remainder in a quarter second, ad infinitum. No matter how long we repeat the process the arrow can never reach the target. There are many variations to these paradoxes, most of them not cited by Zeno (or perhaps cited but not recorded for posterity), but they all amount to the same thing. All involve a foredoomed effort to 'trivialize' an infinity by pointing up its irrationality... not because it is wrong, but because it defies rational appreciation.

The simplest solution to the problem (and therefore one which is regarded as beneath the dignity of learned pedants for lo, these many centuries) sensibly points out that at any given instant we can fix an object in space (a photograph is a case in point), or fix it in speed; which involves fixing a starting point and then timing the interval between point one and a second fixed point. But we can never fix things simultaneously for both position _and_ speed though we can circumvent the problem by saying it was going at a known speed of 60 km/hr when it ran a red light at a specific point. It is just that you cannot determine the speed of an arrow when you photograph it in flight. Note that echoes of this argument appear as elements in Heisenberg's more reconditely phrased indeterminacy principle. Alternatively, I might hazard a mocking sneer and allude to these as finite infinities. "O" is an extreme version of such an infinity. No matter what we do we cannot determine either the precise circumference or area of this dinky little infinity. Pi insists on poking its dirty little nose into the problem so we can never achieve the final decimal point of accuracy. It has neither a beginning nor an ending, its precise contents can never be measured and its circumference can only be approximated. Similarly, Zeno's paradoxes, together with all others premised on the same reasoning, fall into the same category. It would not surprise me in the least to learn that there is an infinity of these finite infinities flitting about like mosquitoes seeking targets to nibble on.

Which brings me to my final nit to pick with infinity; a last hoorah as it were; but the one toward which I was pointing from the onset.

This merely reiterates and flushes out the general theorem I mentioned when discussing the connection between *pi* and the hypotenuse of an inscribed right triangle. It may flush out the reasoning a bit.

I suspect I may have exhausted the patience of my readers with all this esoteric discussion of Cantor and his transfinite mathematics, but it cannot be overlooked when we seek to grasp the enormity of the universe. Those who try to ignore it are foredoomed to fail.

With this out of the way, I shall continue.

IV

A LITTLE LIGHT PLEASE

"The Lord said, 'Let there be light', and there was light!" It sounds so simple. Four little words of power uttered by the master magician of all master magicians.

Ah, but if only it were that simple I wouldn't be writing about it. There are questions every which way when it comes to light, and about the only way to tackle them is one at a time in a more or less systematic manner.

For number one, in speaking of light it doesn't take monumental intellect to accept the fact that we are dealing with a c-set here... maybe; or even merely probably in case there are still doubts.

But first things first, and first of all, "light" is a lot more than 'light'. Light is merely what we call a tiny region of the electromagnetic spectrum which we see when we open our eyes... which implies that to an organism whose eyes are attuned to perceive the deep ultraviolet would not regard our 'light' as light at all and would wind up commiserating with us poor, deprived, blind humans who are incapable of perceiving the universe in all its majestic glory. So light is rather more than just the segment of the spectrum we perceive. Owls, rattlesnakes, bats and other nocturnal animals have 'eyes' that reach well into the infrared, so it is also relative to the equipment employed to do the seeing.

Your reception equipment must address a specific region of the electromagnetic spectrum... which need not coincide with our visual region. In brief, a radio telescope is receiving light just as our human eyes receive it; only we have to go through elaborate receiving and translation machinery in order to 'see' what the radio telescope is seeing. So too does this apply with respect to the electron microscope. All are means to extend the range of our vision.

Light, in the sense we are using it here, comprises probably everything in the universe. There is either light or nothingness (unless nothingness is somethingness)! It is defined in terms of frequencies, which is another term for wavelengths, which range between around 10^{-9} cm (or less) and 10^{11} cm (or more) with occasional gaps which may or may not be significant but likely are there simply because no one has bothered to look for them. In general, throughout this section I prefer to use the word 'light' indiscriminately to emphasize the truth that everything, from ultra-long, 100,000 km waves supposedly left over from the original Big Bang of the universe down to the Compton wave lengths in the subatomic particle range, is basically light. We humans are merely light, but so is the earthworm or the cat or even the bacterium or virus which sickens us... along with rocks and grass and trees which beautify our world. Existing theory promises that everything, from quarks, to virtual particles, down to the tiniest neutrino; including the very atoms from which we derive, is merely light (energy) wandering about with wavelengths somewhere in the 10^{-9} region.

But there are problems here. This is by no means a smoothly running transition. It is not some majestic march descending from the incredibly large to the invisibly small. There are gaps along the way; gaps which range from the trivial (and probably due to a failure to search for occupancy) all the way down the line. Most likely, they fall into the gaps between the architecture of the instruments employed to explore specific regions of the spectrum.

This accounts for the minor problems with the electromagnetic spectrum (or 'light' as I prefer to call it here). The real question mark includes not only the gap immediately before the Compton wave lengths but the Compton wave lengths themselves…which is where the entire range of particles seem to be found.

Some might be inclined to say, "That's all there is; there ain't no more!" And they may well be correct. But they may also be wrong. The jury is still out on the matter.

I do this deliberately, because it may irritate the purists who seek to talk of electrons, protons, neutrons, x-rays, cosmic rays, ultra violet, infra-red, etc. Like it or not, these are all merely way stations along an electromagnetic spectrum of which visible light is the primary method employed by us poor ordinary mortals to look at things.

Another reason for terming it all merely 'light' is, frankly, an effort to rectify one of the more unfortunate obfuscations of the professorial gang. We have the 'electromagnetic spectrum', an x-ray spectrum, the radio spectrum; cosmic ray spectrum, infra-red spectrum, ultra-violet spectrum, and only the gods know how many other spectra lurking on the pages of technical journals around the world. In selecting the 'light' spectrum I hope to jolt readers into realizing that there is only one spectrum…. And everything is on it, and it is everything.

On the other hand, there are also mysteries to be found in this wilderness. One of these mysteries may be seen in the employment of the term *energy*. Are we always talking about the same thing whenever we use it? For a few instances; there is *potential energy, kinetic energy, gravitational energy, solar energy, radio energy, the vim, vigor and vitality of energetic children, latent energy, transient energy, energy seepage, wind and hydroelectric* and likely a dozen or more additional energies, of which *dark energy/matter* is the current favorite.

Can't we narrow the field a bit? Are these all ultimately the same thing? Or are they inherently different?

Part of the answer is simultaneously obvious and subtle. Shall I try it on for size and see if it doesn't simplify matters?

A few centuries ago astronomy was called the "Queen of the Sciences." In the modern world "bookkeeping", with its handmaiden "mathematics" and its master the "computer", comprise the triumvirate "Kings of Kings", with astronomy coming in a poor fourth! Verily, it is the new Trinity, which supplants the tired old ones such as the Hindu Trimurti, the Egyptian Isis, Osiris and Horus, or the Christian Father, Son and Holy Ghost!

Does this sound extreme? Think on it for a minute. Put it in context with the various energies more or less laid out in the last few paragraphs.

This is not science. It is bookkeeping, pure and simple. And the high poobah priesthood performing it are bean-counters with a specialized knowledge of physics and astronomy.

Take for instance, *potential energy*. Hoist a bowling ball to the top of Galileo's Campanile. By the time the ball reaches the top it is bristling with the potential energy it acquired as a product of the energy required to carry it up. When it is dropped it recovers the energy it spent getting up and the books balance.

Now suppose the ball is not dropped. It just hangs around at the top of the tower as a looming menace to pedestrians below.

It remains the same bowling ball it was at the beginning, but it is still pregnant with potential energy. A few tens of thousands of years later it has decayed into a powdery dust atop the tower. But the energy account continues to exist, even though there is nothing left for it to exist on. The books are out of balance... at least not until some bean-counter adds a few bugger factors, i.e. the acceleration of each dust mote molecule as it decays and falls a few centimeters to the deck, the speck of dust whipped up by the wind and deposited the next county over, etc.

Nor is this the end of the matter. When we get out of bed in the morning and stand upright, our head has acquired potential energy. If it should fall off your shoulders it would strike the floor with a thud, thereby exhausting all or most of the potential energy acquired when you stood up.

Not even this is the end of the matter. A reversal is also possible. Dig a well. You expend energy. Try to climb out of that well; which causes you to spend more energy. Now drop a nearby pebble down the well. That pebble had no potential energy before you dug the well, so it acquired that energy out of thin air.

This is nothing but important sounding bookkeeping, calculated by all sorts of bean-counters.

In the same vein are the myriad formal equations which all end in zero, signifying that the books are balanced... the $A+B+C-X-Y-Z=0$ motif so dear to the hearts of bookkeepers everywhere. They ought to call it 'double-entry bookkeeping' and leave it at that.

It works; it is useful, and it is necessary. It is vital to the science of science. But it is not science. It is merely a technique of science, and the technique ought never to supplant the science it is supposed to serve.

'Potential' energy is therefore not truly energy. It comes into existence only when it is exercised, therefore we can cross it too off our list and regard it as a bookkeeper's artifact.

The vim, vigor and vitality of children is merely a descriptive phrase which may be discarded. Latent energy is a whole new ballgame. Consider a rifle cartridge. The powder in the cartridge contains latent energy, which does not express itself until someone pulls the trigger. But if the powder sits around too long the powder deteriorates so the latent energy is dissipated. This is a tough one for bean counters, who scratch their collective heads and recti over it and speak darkly of obscure chemical changes over the years and how the latent energy might be recovered by reformulating the powdery residue... in the process ignoring the energy requirement needed to recombine the mix.

So here we have another bookkeeping artifact. In the case of the bullet we have A+B+C-BANG!=0, but the latent energy did not exist in the real world until the trigger is pulled.

This brings us to a cluster of *descriptive* energies, i.e. solar energy, hydroelectric energy, fossil fuel energy, wind energy, thermodynamic energy, radio energy and those old standbys, kinetic energy and energy seepage.

All these save the last two merely tell us what the source is without saying anything about the character of the energy derived from them. And the energy seepage bit merely says that some of it has leaked out without seeking to categorize it. The books invariably can be balanced by evoking clandestine seepage so let it join the club of phantom energy artifacts. To employ a term from military intelligence, they are merely 'notional', i.e. only real when your foe is deluded into believing they are real and thus a menace to be countered.

So what have we left? We have something called *gravitational energy,* which may or may not be merely a quirk of geometry. We also have an enigmatic phenomenon called *dark matter (or dark energy)*, which may be a bean-counter artifact meant to balance the books. Then there is kinetic energy, of which will have much more to say later.

Finally we have the ordinary, garden variety energy which we see by, or send out on TV waves, or shock us when we stick our fingers in a live socket, power up a flashlight, etc. This energy may be quantified by its wavelength positions on the electromagnetic scale.

There is also one additional energy, which I mention separately because it appears anomalous… magnetic energy. For better than a century during the late 1800's and early 1900's theoreticians sought with grim determination but no success to link gravity with magnetism. The connection seems so obvious i.e. *force* pushes; it does not pull while magnetism and gravity pull not push. Therefore they had to be akin; or so the faithful believed. I suspect that even today there are still those who persist in believing in the connection despite all the evidence to the contrary.

So now we are left with only a scant handful of energies, i.e., ordinary everyday energy, which comes out of a wall socket or perhaps a sun, a gravitational energy which may not actually exist save as a warping of space, kinetic energy and dark matter energy about which little is known although much is theorized.

These are our concern. The artificial energies may safely be left to the bean-counters who need something to solace their wounded vanities. We consider these core energies now.

Galileo demonstrated that kinetic energy equaled one half the mass of a moving object times the square of its velocity, which is expressed as $k=1/2mv^2$. This is a purely empirical relationship. Take any combination of moving matter and multiply one half of its mass by the square of its velocity and you get a number. Nothing exotic or esoteric about that. Give that number a name and a designator letter and you have your completed equation. Call it 'k' and the thing is complete, i.e. $k=1/2mv^2$.

Now we crank Newton's law of action and reaction into the mix and see what happens. The total energy of the transaction is doubled because now you have to take into account the reaction, which equals the action. Now it may be written as $2(1/2m)v^2=e$, where 'e' is the energy produced. Next increase the velocity of impact to equal the speed of light and, voila! We have $e=mc^2$! It is obvious, and it reeks of simplicity. So why did it take nearly four centuries before Einstein came along to put the two pieces together and thus allow astrophysicists, et al. to cackle in glee over the 'latest discovery' of modern science?

Add a little more to the mix. Science still does not know what mass is. We know it resides in matter. We know what it does. We can measure it, weigh it, determine its hardness or softness, and identify its chemical properties and even its structure. We can tear it apart and extract its energy. We can do all these things, but for the most part we remain unconvinced of the reality which says that all that matter is simply solidified light, just as ice is merely solidified water (albeit light is solidified in ways we do not yet fully understand); nor can we hope to understand it until we place it in the proper perspective.   Essentially,   kinetic energy is part of the energy spectrum; but what part? What does this interpretation imply? Is it merely another phlogiston lurking in the background here or is there more to it?

Look to the huge colliders and the particle zoo. What are we doing when we accelerate mass to increasingly higher energies verging on the speed of light and slam them into targets? Forget the hoopla and think about what actually happens.

We accelerate a particle by synchronizing giant electromagnets inside a huge tunnel. This means we are pumping energy into it... just as we pump in air to blow up a balloon. In the process we are adding to its kinetic energy and thus to its effective mass; hence Einstein's "As mass approaches the speed of light it approaches infinity." In effect, the equation becomes $e=c(c^2)$, which is a perfect Cantorian equation that reduces to k=c since squaring or otherwise multiplying infinities leaves infinity.

Now we have a virtually infinite mass. I say 'virtually' because we can never reach infinity. (This will be examined more closely in the next chapter but for now please accept it, at least provisionally, on speculation.) This particle, moving at 99.9999+ c smashes into a target and dutifully explodes, in the process radiating out a cluster of 'elementary' particles which are presumably indicative of the aboriginal composition of the primal atom, or whatever.

(NOTE: This section was written and is inserted into an otherwise completed manuscript in July, 2012 so I can claim no gift of prophecy in discussing its finding of what is believed to be the *Higgs boson*.)

A modest variation on the collision technique has two particles, each pumped to 99.99% c, colliding head on, providing a net combined *kinetic* speed of collision of 199.98 c! The grand super-collider constructed on the Franco-Swiss border was designed for just such a purpose. A virtually immediate development of the collision between two protons was the creation of what is believed to be a Higgs boson. I do not know whether the finding will hold up, but I can accept the theory that something was found precisely where Peter Higgs forecasted we would. It was a brilliant piece of theoretical analysis which deserves our admiration. What bothers me is the background suppositions involved in the discovery and the attendant hoopla surrounding it.

Recapitulating briefly, orthodox cosmological theory keeps reducing the source of our universe to ever smaller dimensions. Currently the start-up size is postulated to be less than 0,001 cm. Within that fraction of a cm is packed all of the energy contained within this universe... all of the galaxies, all of the individual stars and planets and all else wherever or whatever it may be.

Why is the size so reduced? Simply stated, the mass of the universe clearly exceeds every available explosive force. Even a super hyper nova could not expel this mass and eject it out into space, ergo, since the aggregate mass of the universe is not negotiable we must seek out some mechanism of increasing the effectiveness of the explosion.

So far so good. Successive reductions in the dimensions of the start-up size of the universe helped, but not enough. Additional theories were required to make it work. How about throwing in an additional dimension? An explosion which temporarily nullified the wall separating two universes might do the trick.

In the 1930's P.A.M. Dirac proposed a duality where the dark spaces of our universe were bright regions in our companion universe while the reverse was true in the opposing one. The idea never gained much traction and I suspect Dirac had his tongue firmly in his cheek when he proposed it, but nonetheless, it was the first theory to my knowledge which advanced the concept of multiple dimensions to account for our universe.

But what force might be capable of creating the rupture? Nothing in our books was up to the task. Until 1964, particle scientists were thoroughly at sea. Nothing seemed capable of generating the necessary force. Then a trio headed by Peter Higgs of Edinburgh University postulated the existence of a super-particle capable of injecting mass into 'massless' particles, such as electrons and other, normally 'massless' fermions. These carriers of mass were collectively called 'bosons', and they came from the other dimension and attached themselves to stray fermions, thus becoming bosons capable of uniting into atoms and molecules to create stars, planets, galaxies, etc, and ultimately humans capable of dreaming up dimensions and bosons.

Interestingly, this still did not explain what mass is. To this day we still don't know, and if the Higgs model is correct we may never know. It will remain forever an alien mystery. The whole affair is pretty convoluted, but at least it provided a somewhat rickety escape from the dilemma of accounting for the big bang.

There is more to be said on this matter, but for the nonce I intend to ignore the implications inherent in the Higgs' boson argument until I can treat it in context of my declared aim of creating a working framework for cosmological exploration; one which does not require outright mysticism.

First, suppose we take a look at what these giant colliders are doing; only this time we start from the logical beginning rather than some arbitrary human jumping-off point. Do this and a different perspective emerges.

We take a proton and run it through an array of magnets, each pulsing at a carefully timed rate which causes the proton to pick up speed. Either of two techniques can be used to achieve this, i.e. a circular 'cyclotron' or a lineal accelerator. Either of these techniques can achieve accelerations to speeds running around 98 or 99% of light. Tweaking the power output of the magnets can achieve speeds of 99.99 c or so. The imparted energy being less than c, it is evident that the ejected proton cannot quite reach c, but it approaches it very nearly. When it reaches the maximum speed the collider is capable of delivering it is smashed into a target and fragmented into a myriad pieces.

It is also evident that no fragments of the collision can exceed c. This is a built in limit since nothing can exceed c. All we can do is add a few more 9's to the right of the decimal, perhaps to 99.9999+% [We will get to the matter of light speed in the next chapter. At the moment I am staying strictly within the terminology and ideology of orthodoxy].

Colliders offer a method of examining the effects of collisions verging on twice the speed of light. Here we are not accelerating a proton to strike a stationary target. Instead we are accelerating *two particles* in opposite directions and photographing them at the instant of collision. The Franco/Swiss Super Collider thus achieves a combined velocity of 199.9998% c and a mass only slightly less than a Cantorian mass/infinity! Studying the ejecta revealed an excruciatingly brief flicker of a particle which seemed to fit the Higgs boson model. Conventional photographic plates would never have caught it, but an array of electronic cameras caught it before it blinked out of existence.

So far so good. It all seems enchantingly impressive. But there is another interpretation available... one pretty much forced by my analysis to this point.

Visualize a solid block of glass placed beneath a pile driver. Surrounding this glass at a safe distance are arrays of cameras and other instruments which together capture the event at the instant the pile driver smashes into the block.

As the pile driver crushes the block of glass the latter vanishes into a heap of silicon dust which is not captured by the detection apparatus since it has effectively reentered the background universe in the form of dust. But shards and slivers of glass jet out the flanks of the pile driver at varying speeds and all sorts of trajectories, all of which are duly captured by the detection equipment to be examined by physicists who are seeking to unlock the secrets of universal glass blockdom by their findings. The experiment is repeated scores or hundreds of times, varying the types of glass and the momentums of the pile driver.

The net result of these experiments, using various collisional velocities and target substances leads to bewildering results and increasingly outlandish theories, often involving hypothetical 'virtual' particles and miscellaneous, rather mysterious, forces in order to explain their findings. But surprisingly, each additional 'discovery' leads to a new mystery. Around and around it goes; and where it stops nobody knows.

The matter is simple; pump propelling energy into an object and then fiddle around with it a bit. Throw it at an ordinary bathroom scale tacked onto a wall and shoot a bowling ball at it. When the ball strikes the scale the scale obediently registers the strength of the impact which, low and behold, somehow contrives to equal $1/2mv^2$ as the end result. Without ever intending to do so you have matched Galilio! What happens? You can merely hope to recover the energy you pumped into it, possibly with some of the already existent energy originally contained in the object... which does not affect the equation since it now becomes necessary to crank the mass of the target object into the calculation.

So kinetic energy is not necessarily part of the energy spectrum and is instead simple light. Instead it is merely a descriptive word, a notional word, intended to distinguish a separate application of energy under specified conditions. It is not an entity unto itself. Returning to our train of logic, we have now just the main spectrum of energy to consider, as well as gravitational and dark energy. As our primary tool we have only the electromagnetic spectrum to play with. If we can assign all of these energies to places on this spectrum then we may safely declare that the universe consists of nothing save energy and nothingness: with the nothingness unreal unless it is somethingness. So I propose to steam full speed ahead and damn the torpedoes, with the next goal being to consider the nature of magnetism; which has long been something of a problem.

At the exceedingly short and long wavelengths it is an open question whether there is more to be learned and if it is of any importance if we find it. In anticipation, I confess to a belief that both extremes are highly significant, but for the nonce I mean to wait until I get to the appropriate place to address that question.

This said, let us move to a reasonable starting point, and begin at the upper range of the visible (to us) spectrum, where the wave lengths are well defined and pretty much explored. Consider the speed of light; the "c" in Einstein's mystic equation as well as the veritable epitome of Cantor's 'c-sets'. Efforts to measure the speed of light hark back to the early 1800's. The concept employed was commendable and the technique was high standard for the era. There is little point in going into detail. Over the years it has been refined and newer techniques for determining it have been developed. By now it is merely a matter of adding or subtracting decimal points. But the Michelson-Morley and following measurements brought the speed estimates up to the vicinity of 299,793 km/sec where it has fluctuated modestly ever since. I doubt if any two observations or calculations precisely agree with each other, but the differences under properly controlled conditions agree to within a few kilometer/seconds or so of the 300,000 mark. A difference of 207 km/sec is so trivial that in general in general we can call it 300,000 with reasonable confidence we are near the mark.

Problem solved!

Or is it?

Still no one seemed to know whether light is a wave or a particle. Sometimes it behaves like a wave, but on other occasions it behaves as if it were just another particle. It lacks any detectable mass, but strange to say, it exerts a quite easily measurable pressure when it strikes molecules such as gas!

This was in itself a major stumbling block. There is nothing too surprising at the finding. Nature has a habit of blurring the edges of many sorts of categories. The boundary between mammals and dinosaurs is blurred by the duck-billed platypus, among others. It is a mammal since it has hair and suckles its young, but it has poison spurs on its heels, lips like a duck, hips that displace the legs out to sides like reptiles a tail like a beaver, is monocloacal and it lays eggs! Anatomically it possesses features similar to those of the great saurids of the dinosaur era, plus the descendants of dinosaurs, or 'birds' as we call them today. In addition to avians and reptiles it also possesses mammalian anatomy to boot.

All in all, it is a thoroughly confused animal; either that or God has an antic sense of humor together with a penchant for practical jokes!

(Personally, I much prefer the idea of a laughing God than the current widely perceived idea of a humorless solitary deity flailing about angrily condemning souls to Hades for failing to conform to the prejudices of priests, preachers and assorted ayatollahs, all of whom forever seem to be ordering their God about like a puppy on a leash, commanding it to smite anyone who fails to succumb to their prejudices, their hates, or their lust for power and authority.)

But I digress.

Nor is this all. In the microscopic world of nature there is blurring between the animal and vegetable kingdoms, etc. So why should there not be a comparable blurring between waves and particles? Or perhaps we have become too obsessed with an 'either or' approach to the universe?

There are several very good reasons to be concerned when it comes to light. Recall Galileo's formula for determining kinetic energy. $K=0.5mv^2$ (kinetic energy = half the mass times the square of the velocity). But according to the rules of mathematics if you multiply by zero the result is always zero! So if light has zero mass it must have zero kinetic energy and thus can exert no pressure, and a beam of light that can switch from a wave to a particle and back again as the mood strikes it is enough to give the stodgiest physics professor a few gray hairs! It is also enough to send a myriad mathematics professors scrambling to come up with some equation (any equation that works) which might conceivably be applied to conceal this defect in theory.

This does not presuppose that there is actually a real defect in the theory, but it surely ought to be enough to make us suspicious. Whenever I run into a conundrum such as this my first inclination is to go back over the initial formulation to see whether a mistake may have been made *ab ovum* (or gut error to those oblivious to puns). So suppose we take another look at the $k=1/2mv^2$ equation. Perhaps later findings may have modified Galileo's findings.

Alas, 'tis not to be. The experiment was absolutely straightforward and the results were equally straightforward. Worst of all, over the years any number of high school and college students have replicated the experiment, and if even they cannot arrive at a myriad different solutions I must believe that the original experiment was a smashing success.

Eventually we got around this obstacle by recalling that there is no law of nature requiring it to obey human laws and in this case light has been assigned a pro forma mass which just happens to be calculated by measuring its effect on gases and particles in the laboratory and elsewhere. When we do this..... surprise, surprise! It works. We actually come up with a usable number, as well as ex post facto explanations of why it works... all of which boil down to saying, "Look, we've calculated it and we've encrusted it within a blizzard of mathematics aimed at showing how it works, so sit down, accept it and shut up!"

Maybe so. It is a classic example of the tire-patch approach to physics, and it remains a clunky solution. But for the nonce I am concerned with the question of the velocity of light and the numbers do work out so I accept them, though I remain in doubt of the interpretation most commonly advanced. The questions of mass and gravity should be addressed somewhere in this because they are a critical element in both the problem and the solution.

Why should this be, you may ask? That comes in the next chapter, more or less, depending on what we learn first.

V

2C AND/OR not 2C 'PANDORA'S BOX'

Definition: We are accustomed to using the term "photon" more or less as the smallest unit of light. This is a dangerous presupposition; particularly since it implies a connection with the wavelength of light, which is merely a small segment of the electromagnetic spectrum. I can accept the idea that a photon may have a variable length... perhaps enough to transcend the visible, or recoverable, portion of its spectrum, but we have detected ultra-long spectral waves supposedly relict of the Big Bang, where the wavelength is around 100,000 km (one-third of a light second). I find it impossible to believe that a solitary photon can be stretched this far... (Or even 50 or 100km); or more properly, that a solitary minimum unit of light can be thus extended. This absolute minimum of light, call it the *reduction ad infinitum*, I term a 'phote' so wherever the term is used this is what I mean. I posit it as the minimal particle (or wavicle) of light and one which is independent of the technique used to determine the wave length of a photon.

In short, a photon is postulated as consisting of a linked chain of photes traveling just as a wave in the ocean travels, i.e. as a wave of molecules of H_2O linked together in an apparently coherent chain.

Carrying the analogy of oceanic waves to starlight further and we may accept the concept of three dimensional bundling and visualize a sort of a sheaf of light traveling in company with adjoining sheaves sharing the same wave length. This seems to be a trifle ornate, but that is not necessarily a fatal objection so I leave it at that and mean to return to it later should the need arise.

Accept the fact that 'c' denotes an infinite speed. (I have to because I follow Cantor's lead and thus have defined it as that.) Since everyone else in the physics trade pretty much agrees, that closes the matter. Nothing esoteric about it. I am called 'George' because my parents named me George. 'C' is called infinite speed because we have named it that.

I recall a time, roughly 60 years ago, being taken to task for saying that the speed of light might not exactly equal c; that there may be a 'C' for infinity and a somewhat slower '>c' pertaining to the speed of light. But the question of 'c' is the linchpin for virtually all of modern physics. It is also a major conundrum; one sometimes best approached via 'black box' experiments. But here comes another of my "But firsts" so you may as well resign yourselves to them. This 'but first' is pretty prosaic and otherwise unremarkable.

According to current theory, if light is utterly massless then the least nudge by a body possessing any degree of mass will send it darting madly off into the empyrean at infinite speed. This is merely a logical extension of the bowling ball striking the ping pong ball experiment, where the disproportion between the colliding masses leads to a huge recoil by the ping pong ball, but a virtually zero one by the bowling ball. So there is nothing terribly exotic in the conclusion that a zero mass will be accelerated to infinite speed if impacted by an object possessing any mass, no matter how small. But there is a sneaky little codicil tucked into this conclusion; one which should be mentioned now before it leads to misinterpretation. If we postulate some element of mass in light photes then Newton's equal and opposite reaction, as well as the whole inertial resistance thing, must enter into the picture so an infinite speed can never be achieved! No alternative exists.

You ask why I can say this with so much certainty. Think on the matter. As stated, the speed of light is necessarily >c and thus somewhat less than C. Therefore, some component of its energy must be subject to the impact/recoil effect and thus cannot partake of any acceleration to c. The Inertial resistance of mass also must be taken into account.

So let us see where that leaves us.

We start by supposing that light is truly massless and >c therefore equals c and find out what this implies?

First consequence: According to Einstein, as mass approaches the speed of light time approaches zero. Though I decline to parrot any authority without first having studied the logic underlying it with a critical eye, this dictum has been pretty well verified by ultra-precise atomic clocks subjected to orbital speeds around Earth. They have revealed an exceedingly slight slowing in clock speed commensurate with the increased speed of the satellite. We may therefore accept, at least conditionally, Einstein's conclusions here. But we must do so with appropriate caution; among which is an absolute need to abandon our parochial attitudes and adopt a more universal standard where all the myriad vectors of motions imposed by an active universe are taken into account.

And this must be impossible to achieve simply because there is no conceivable way of completing the task.

It may still be possible to dispute these findings; not altogether likely, but certainly possible. So suppose we try reverting to ordinary logic and look around to see where a touch of rational thinking leads us.

For the nonce I am primarily interested in the element of time in relationship to the speed of light. Nor is this an idle interest on my part. No theory of everything can hope to be successful if it continues to start midway; at the point of some possible, or probable, "Big Bang". It must start at the beginning and go from there. Entirely too many alternatives are left dangling in midair (or mid-space) if we opt for the midway start approach.

Extending the consequence, if time slows for mass as it nears the speed of light then there can be no internal time for zero mass phenomena moving at the speed of light. To suppose otherwise we would ultimately be compelled to conclude that some wholly unexpected (and unexplained) capability of zero masses can literally suck the phenomenon of time out of massive objects.

By extension, truly massless light must exist in a never-never land where a solitary 'phote' can travel from one verge of the universe to the other in zero internal time, and thus without any alterations or basic change. There is not even an instant when a sense of transition has occurred. As a far-fetched analogy, if a traveller aboard our hypothetical phote had just taken a bite from an ice cream cone and at the precise instant he bit into the cone, he transitioned into light speed, then when his vehicle arrived at the far rim of the universe and dropped back to normal speed, billions of years would have elapsed on the outside but the ice cream cone would not even have begun to drip!

I do not apologize for belaboring this subject simply because it is so far removed from human experience as to provoke involuntary disbelief. Still, this is precisely what Einstein was getting at when he enunciated his Theory of Relativity, i.e. there are inevitably many frames of reference for the same phenomenon and all of them are legitimate points of view.

A subtle potential consequence of this conception of time addresses the matter of the wavelengths of light. If light is actually massless there can be no internal clock and the only way by which light can lose wavelength and relax is through collisions of some sort. But if light possesses the least amount of mass then any fraction of its wavelength imposed from outside is free to relax, thereby increasing its wave length. By extension, if the internal clock ticks even once during a journey across the universe, there will be a noteworthy reddening of the spectrum; and the possibility that a total decay of the wavelength will occur.

So I try again. Take time to think this one through. Consider what is said here: A phote moving at the speed of light experiences no sense of time. If we place our disembodied minds atop a single phote at the nethermost verge of the universe and ride it to the opposite verge we discover that the subjective time lapse in traveling between the two extremes is precisely zero! It is instantaneous so far as we are concerned and all the light which impinged upon it during its journey is merely a part of ambient luminosity created by the passage of an infinity of photes going every which way through otherwise empty space.

And this makes it time for another digression.

Modern cosmology is saturated with space. There is a "fabric of space", a "warped space", a "folded space", etc. There is also a "property of space", as if it owned real estate or other assets. But it seems as if there is no such thing as "empty space". Which is something of a shame. You'd think that with all these 'spaces' floating about someone would have a clear idea of what he is talking about. So suppose we step back and start actually thinking of what we have in mind.

Just about everyone has seen the analogy involving a solid little sphere dropped onto a stretched rubber sheet. Einstein started the ball rolling when he pointed out that the profile of the sheet with the ball resting upon it is rigidly mathematically identical to that of Newton's inverse square depiction of planetary orbits....which it is. So Einstein correctly stated that it is convenient to regard gravity as a simple inertial field created by the presence of a massive object stretching the fabric of space ----- which indeed it may be; from which we may legitimately infer a warping of space, and thus a reality of space as opposed to a merely notionally abstraction about space; which may be a trifle premature.

At any rate, the physics chaps quickly picked up on the new gospel and ran with it. So now we have a fabric of space to play with. But it is a flawed argument since it is purely two dimensional and the relationship between a planet and the space around it is three dimensional.

There is also old Euclid to consider. He too had something to say on the matter well over 2,400 years ago! According to him if you take a circle and project a pair of radii at an acute angle at any arbitrary distance from the origin, then draw another arc at twice that distance the second arc will necessarily be twice the length of the first. This process may be extended to infinity, with each doubling increasing the length of the arc by a factor of two. In three dimensional space this means the attractional force of gravity diminishes as the square of the distance. The resultant diminution of the gravitational field therefore precisely duplicates the 'stretched fabric' analogy presented by Einstein. And it does so without any requirement for a 'fabric' of space!

So why bother postulating a 'fabric' of space in the first place?

Einstein was the one at fault here. He was thinking in two dimensions when the object being considered must be perceived as three dimensional. When we add a third dimension to the picture to depict a solid body such as a planet or star and repeat the process. Guess what? We discover the same relationship holds true.

In other words, we do not need a warping of space to account for the phenomenon of gravity! It may be that this is the case and thus there is indeed a warping of space, but it can by no means be proven by this sort of observation. Einstein's employment of the inertial field analogy, with its codicil inferring a "warpable space" is at best a suggestion, not a dictum. It is merely a useful analogy, one which is not only misleading but has created a new generation of true believers to convert the analogy into an article of faith...a Holy Sacrament of physics, fully bathed by the sweat of the physics chaps who play with it!

So we are back to square one. Space is either warpable... or it is not. 'You pays your money and takes your pick'.

Shall we try another tack?

Look more closely at the stretched sheet analogy, this time from the standpoint of a three dimensional field. Now we are confronted by a whole new perspective; one which proves rather disturbing.

We have a massive sphere which we may call 'Earth', engulfed within a warpable space and our cute little stretched space analogy becomes a sphere of continuously lessening emptier space surrounding it.

How to interpret this?

I suppose one enthusiast could create a mathematical equation to 'prove' that the Earth is somehow drawing space into it while another might construct an equally beautiful equation which 'proves' that space is sucking some of Earth's gravity (or other vital essence) from it; or perhaps a third might construct a mathematics based on the premise that Earth and other massive bodies are somehow exerting a repulsive force on the fabric of space, thereby warping it!

Sarcasm aside, the reality remains: None of these provides so much as a hint of actual evidence to prove the existence of a 'fabric' of space! Nor do they do anything to disprove it, though this does suggest that it may actually be wholly non-existent. At the very least, if Ockham's Razor has not been repealed, and remains a canon of science and logic, there appears to be no coherent reason for 'warped' or distorted, uneven, or tattered space to be taken as anything more than a purely notional product of "voodoo science".

A closer look at the internal logic of a 'fabric' of space makes it even more disturbing.

We have a universe consisting of nothing save permutations of light; with odds and ends of alien matter called 'mass' possibly thrown in for good measure. Next we toss an outside something/nothing which we call space into the mix. But this light somehow interacts with this nothing to distort the even distribution of the 'nothing' without otherwise altering it. To make our observations even more arcane, this 'nothing' which is distorted by our light manages to do 'nothing' to the mass which distorted it. The latter continues diminishing in effect according to the same inverse square rule predicted by the ordinary, everyday geometry formulated by Euclid more than 2,000 years ago. There are other arguments which could be thrown into the pot, all of them arguing against the 'fabric' of space theory, but perhaps the most convincing one stems from the simple fact that to warp space that space must be tightly woven in order that it may be warped evenly throughout and thus embody the stretched sheet image. But nowhere, except in the human imagination and our human mathematical procedures do we find the least iota of evidence for this.

Summarizing, the idea of warped space has become a dumping ground for modern astrophysics. Anything we don't understand we casually dismiss as a merely another 'property' of space. In short, all the warping, folding, smearing or whatever of space may well turn out to be the 'phlogiston' of the 20th Century. May I even suggest that we ought to start alluding to the "phlogistic space' to go along with its other attributes?

Returning to our chain of logic prior to my latest digression, the fact that an observer riding piggy-back upon a massless phote does not experience time does not mean that an observer tracking a specific phote from an outside platform would see simultaneity at every point; nor even from any two adjacent points. As an outsider I might pinpoint a beam of light passing point (a) at moment (x), passing point (b) at moment (y), and point (c) at moment (z). This would be a characteristic signature of a finite observation system. We might perceive this light as having traveled billions of years, but from the standpoint of the phote itself there is not so much as a flicker of elapsed time.

This is precisely analogous to carving out a section of an infinite line and dividing it into meters and kilometers. We may slice arbitrary segments of time and convert them into distances without affecting the infinity of time. 'Time', as a term, is therefore meaningless to a phote and is applicable only to phenomena moving at sub-'c' speeds. Additionally, the faster the phote is moving the slower the elapsed time between any two points as seen by an outside observer, though it would be meaningless to the phote itself.

Simplifying, we watch an automobile speeding between point A and point B. If it is moving at a moderate 100 Km/hr it takes an hour of time to get there. Increase the speed to 200 km/hr and it gets there in half the time; ergo. As the speed increases the time it takes grows shorter. Now that wasn't so hard, was it. All the fancy talk boils down to this small snippet of logic.

Several notes of caution are called for here. Most significant is the role of the shortest wavelengths of light; those running around 10^{-9} cm and less. This is the region where our finite universe enters into the picture. Mass is undoubted, even if we don't know what it is. Gravity is obvious, even if we don't know what that is either. Odd little critters, such as virtual particles, electrons, protons, neutrons, neutrinos, atoms, and ultimately even bulky objects such as us humans, whales and elephants; can only exist through the consumption of light in the form of food, air and water by utilizing light as nutrition. After all, there is nothing else in the universe so my statement must be true... unless someone somehow comes up with a technique for extracting nutrition from phlogistic space; which seems *a priori rather* unlikely! What we discover in this ultra-short region of the spectrum is mainly a rather disorganized assembly of embodied wavelengths usually called 'Compton' waves in honor of their discoverer.

But we still do not know why! What is the mystique underlying these minuscule wavelengths?

Flitting from one digression to another, I would be remiss in my responsibilities as a man, and prove myself an intellectual coward, if I failed to remark the theological implications of this focus on light. As stated, light is the solitary occupant of the universe. Lacking it there is only void ... boundless emptiness save perhaps for alien mass.

Light is indestructible, immortal, omnipotent, and omnipresent. Thought can only be achieved through the instrumentation of this light. The very process of thinking is epitomized by light, and anyone who has read the comic strips in newspapers recalls the image of light bulbs flashing on to indicate the onset of an idea.

Light creates its own laws and can never be denied. In short, in the ultimate sense, light is the only thing that is. As such, it fulfills every one of the requirements traditionally attributed to a deity. It is the source of all things and all phenomena. The Biblical mandate of God which ordains "Let there be light , and there was light." was closer to the mark than its authors ever imagined! And the Zoroastrians, Hindus and Buddhists may have been correct all along!

It is an interesting possibility; one which may mandate a deeper study than I am capable of giving.

Does this mean I have penetrated to solve the ultimate mystery of the universe and proved that a God must exist? Is it possible I may have uncovered the actual persona of God?

I don't know about that. It all depends on what we mean by the term "God". Are we speaking of a self-aware, rational entity? If so, then I have proved nothing. Nothing I have said or described here contains any implications of possessing an active self-awareness or suggestion of a rational purpose. But neither does it exclude the idea.

Nor have I described anything truly new in this. The oldest original religions, among them the Hinayana (or 'little basket') branch of ancient Buddhism (ca. b.c.400), portrays precisely this condition for its God, with the ultimate goal of mankind being to free themselves of the delusion of existence and a return to the passionless, emotionless, thoughtlessness of the Nirvana. The Hindu 'Bhagavad-Gita' and the Zoroastrian 'Zend Avesta', both of which developed far earlier than Buddhism, as well as several Greek and Roman philosophers portrayed the gods as entities totally disinterested in the affairs of humanity or any aspect of this world. They started it, then washed their hands of the whole affair.

Even in the Judeo-Christian account God is reported as having repented of creating man and several times threatened to demolish the whole experiment ... which is not much different from abandoning the experiment and walking away from it in disgust.

In effect, from what I have said thus far, the verdict must be given over to a Scot's jury and rendered as a definite "Not proved". These observations and conclusions are in no way theological basis for any conclusions of faith. They are purely and simply rational deductions which attend the investigation of infinity and the solitude of light as the only occupant of the universe.

Any theological conclusions are beyond the purview of science and remain in the realm of speculation and personal faith. The solitary theological position supported by my digression might be agnosticism, where atheism is acknowledged as being merely another religious faith. The statement that "The theologians have failed to prove that there is a god, therefore there is no god!" is precisely as irrational as the contrary statement that "The atheists have failed to prove that there is no god, therefore a god exists!" Both are expressions of faith which lack any connection with logic or proof.

The solitary wholly rational position is neutral. There is no proof either way. Perhaps even more importantly than this, there is nothing we can do about it even if we did know. So we might as well relax and accept whatever may come. And that, my friends, is purely agnosticism.

Enough said on that score. It all sounds very impressive, but there are a number of problems with the scenario laid out here. Most significant, perhaps, is the acknowledgement that light, at the Compton wave lengths, obviously possesses mass, and therefore gravity. If it didn't I wouldn't be here to talk about It; nor would you be here to read it. It also appears to stand apart from the remainder of the spectrum since the largest gap in recovering wave lengths falls in the region immediately above the Compton lengths.

This is troublesome on several levels. It leaves us with a total of three alternatives, only one of which seems even modestly comfortable.

The first, and seemingly most probable explanation is that every phote of light possesses a minuscule mass and gravity. Being this tiny our instruments simply are too gross to work with them, and trying to do so would be rather like using a sledgehammer to drive a tack. Pack enough photes together in one place and we get measurable mass, matter, and gravity as byproducts, but lacing a sufficient quantity the mass or gravity is undetectable.

This has the advantage of fixing mass and gravity firmly on the electromagnetic spectrum. We continue to have only a solitary phenomenon in the universe, albeit fortunately as a phenomenon which is singularly mutable in its manifestations. It is, as I see it, the most likely solution to the problem.

On the negative side we have the fact that we have yet to detect any real proof of this and efforts to unearth a measurable mass in photons have repeatedly come up empty. But this failure is rather negated by the fact that the gravitational constant works out to be beyond the reach of most experiments, so our conclusion is by no means refuted, though it must be in some doubt pending future developments.

The second, and I think considerably less probable alternative, calls for an outside component, say perhaps on the same ratio of sizes as a virus to a bacterium, which can interact with the shortest wavelengths of light to impart mass and gravity to objects on this scale. A priori this seems to be a rather unlikely solution, but the fact that we find something of the same order as viruses at the smallest extreme of the 'life' spectrum, as well as the neutrino in subatomic physics, must give us pause. After all, the existence of viruses was not discovered until the late 1930's, so it is not unlikely that many more surprises await the next few generations of research technology.

I am convinced that whenever a gaggle of experts declaims from on high that a phenomenon within their specialty, cannot be explained in terms of an analogy culled from another specialty, they are treading on exceedingly fragile ground. All too often nature seems to work on the analogy principle so I accord a degree of plausibility to this approach even as I strongly doubt if it applies here.

The third option is superficially perhaps the simplest of all. It merely asserts that energy at Compton lengths is not the same as other types of energy.... That there are two or more separate and distinct phenomena mislabeled or misthought of as a single thing. This would mean that there are several different types of energy, inherently separate and entirely distinct from one another. We might regard the present approach, where all energy is thought of as being the same, as the equivalent of declaring baboons, chimpanzees and mankind to be merely humans. This would not necessarily be incorrect even though the chimpanzees might object to being classed alongside in the same category as politicians, ayatollahs and other self-appointed spokesmen for God.

Despite its simplicity I regard this as having only an outside likelihood of being the answer. One might assume that if there is a true or significant difference between two different types of energy our ever-intensifying experience with accelerators would at least have given us a hint of such a difference. And thus far we have discovered no indications of such a thing, though the physics chaps have identified certain bosons as carriers of mass, and therefore of gravity. Of interest here is the explicit specification that mass and possibly gravity are visitors from an alien universe in a different dimension --- which I suspect is another bit of phlogistic mysticism!

I should append a brief note here on the fact that magnetism does not seem to slip comfortably within the electromagnetic spectrum; and this despite the fact that it is clearly electromagnetic in source and origin. I mean to return to this little tidbit in a later chapter, but in the interim I urge you to keep the problem in the back of your minds. It is a significant element in cosmology.

I know I cited a total of only three alternatives here, but several years ago I focused on a fourth one and devoted six years to it before dismissing the idea for lack of pertinent empiric data to sustain it.

It is quite convoluted and to a degree is dependent on the results of the findings with respect to my original three alternatives, but it contained several interesting avenues and offered explanations that were lacking in conventional models. Its main flaw, so far as I was concerned, was a requirement for two fundamental forces in the universe, i.e., amorphous energy possibly consistent with gluons, and helical gravitons. Between them, they made up the universe, including matter.

It was a clunky approach. Worse yet, it was one which was aesthetically unpleasing, but in general it worked. I would not now be mentioning it save for a tiny newspaper announcement In March, 2010, which stated that the dark matter of the universe had been found to obey the rules of flat geometry!

In Cantorian terms this would mean that where Einstein derived a universal geometry premised on a Cantorian set requiring a curved infinity, here we uncover a fundamental system working on a Cantorian straight line infinity!

Since the problem of geometry was the great sticking point which led to my dismissal of it as an alternative, this fourth alternative found itself suddenly snapped back to center stage! Is it indeed possible that there are two different types of energy wandering about in the universe?

Consider the situation. We have an excellent operational definition of mass; i.e. we can measure it on a set of scales, we can determine its momentum and its inertia. We can study its dimensions. We can play all sorts of mathematical games with it. We exist because of it. But despite all this we still do not know what mass is!

Now consider the fact that atoms seemingly consist of various combinations of ordinary energy. But a new ingredient suddenly appears somewhere in this mix. We do not know what it is, but we know what it does. Some of the physics types confidently go back to the particle physics consequent to their accelerators and tell us that this or those particles are bosons and the "carriers of mass".

But what is this mass they are toting about on their backs, and where did it come from? Ask this question and all is silence. They haven't the foggiest idea. Mass is something we can measure, but we do not know what we are measuring. The best they can manage is a soft mutter that perhaps it comes from an alien universe off in some different dimension.

Much the same may be said of gravity. We have measured its constant of value (in terms of what it does to mass. We study its effects through systematic study of orbiting astronomical bodies. We employ it with great success to calculate the trajectories of exploratory probes aimed at the very fringes of the solar system and beyond; but the best we can manage in accounting for it is to postulate another phlogiston in the form of a mystical distortion of space!

Clearly, there is a fundamental gap in our understanding here.

So suppose we start by taking a closer look at mass and see where it may lead us.

I propose that one of the flaws in modern physics is its stubborn determination to compartmentalize everything. Someone sees the track of a new particle on a photo plate and announces that it is the product of such and such an operation and belongs to such and such a family of particles, some of whom are carriers of mass. There the matter drops and we go on to the next particle.

But why stop there? One might logically poke around a bit deeper and start looking to see what else this particle might do! Where is the law of physics which requires that every particle do just one thing, and one thing only? Lest I leave you in suspense here, yes, I regard this as the real solution to the problem and can pretty much guarantee it will come as a surprise, mainly because it has never, to my knowledge, even been hinted at by any reputable physicist .... Which I suppose marks me as an utterly disreputable physicist.... which, on the final analysis, may well be true. But even disreputable, I am still original and willing to think outside that fabled box, which so often winds up converting itself into a coffin lacking in all save the stalest of ideas. So here I go and damn the torpedoes!

I know I have not answered the "C" or >C question here, but that was never my intention. All I wished to do was to lay out the alternatives and investigate the consequences of each. The final answers will be offered later. Until then, I proceed.

VI

LIGHT IN DARK PLACES

By now I suspect many of my readers (assuming I have any left) are wondering why I pay so much attention to the possibility that >C does not equal C. Why fret about esoteric Cantorian infinities? Why is the question of whether light should be assigned to the set of all curved lines rather than the set of straight lines important? And, perhaps the most significant of all, how can the infinite relate to us finite critters?

I have no doubt that I am pretty much a minority of one in believing this, but I am convinced that no successful theory of everything can be launched by beginning in the middle and meandering along the byways of physics; which is precisely what we have been up to for more than a century. That approach begins with the assumption that everything must start with us humans. What began before us is history. What comes next is future. To us, 'light' is the minuscule wave length we see by. All else is a thought of as a separate study. As an abstract proposition we may accept the proposition that everything is merely light, but our subconscious minds continue to insist that there is an impassable gulf separating a human body from a beam of light; regardless of the "Beam me up Scotty" mantra of the Star Trek sf shows.

This intellectual quirk has created an unfortunate mental block which subtly warps much of our data processing. To illustrate, our giant accelerators and the physics practitioners manipulating the apparatus of subatomic physics never seem to get around to addressing the question of how these things relate to one another. Some physicist finds a new particle lurking in the collision spray in an accelerator and the announcement goes out that it is a carrier of mass. Everyone is satisfied and no one ever stops to ask what else this particle might also do.

All this is but the tip of the iceberg. Poke around beneath the surface of modern physics and we see enormous progress in the creation of mechanical tools aimed at refining our existing knowledge or ruling out existing theory, but virtually nothing in the way of poking around outside the box to see what else may be lurking in the wings.

Expanding on this thought, some workable protocol must outline specific avenues to be pursued and the starting point for a workable theory of everything should begin with the broadest possible understanding of light and then go from there.

My reasons for saying this will become obvious here in Part 2. So please bear with me as I meander down unfamiliar byways and watch where I emerge ....assuming I don't get lost somewhere along the way and somehow drift off into the sunset.

Cast a stone into still water and you get a circle of ripples extending out from the point of entry. Listen to the mournful howl of a steam locomotive as it approaches, then passes, changing pitch all the while as a Doppler effect. Study the spectrum of a planet as it orbits its sun and note the manner in which it shifts from the blue end of the spectrum to the red end as it varies from approach to retreat.

The idea of wave lengths and Doppler effects is so commonplace as to be virtually trite. But there is an anomaly here if we stop to think of it.

Where did this phenomenon originate? Conventional wisdom says, 'why ask? That's just the way things work. Light has a wavelength and that's it.' And conventional wisdom may be right... but then it may also be wrong. If no one asks an unconventional question no one will ever know, will they?

So we peer at a nascent universe consisting of nothing save a vast ocean of flat, dark energy lacking any hint of structure or wave linkage. It may help to think of it as a pervasive, London type pea soup fog. Time does not exist, nor does any sense of motion. In the ultimate sense it must be dimensionless, timeless and wholly blank. Since it is dimensionless and timeless, it may as well be defined in our minds as smaller than the head of a pin or, alternatively, as spanning an unimaginably vast expanse of space. Since there is no time, there is no way of calculating duration, no sequence of events to provide a starting point or base to calculate from.

We may regard it as either a megaverse or a microverse with equal validity, but to accommodate our human imagination we may as well think of it as a megaverse since it is easier to squeeze more stuff into it than into something we regard as minuscule. But this is merely a convenience and need not be taken literally.

How many successive apparitions of this megaverse may have recurred before the present one must remain unknown. The one we are in may be the first, but that is iffy. No matter whether ours is the first or the thousandth, there was an initial instant when this is all there was. So how did wave lengths emerge from this soup?

Nothing in the physics texts offers a solution to the question so it remains as an unasked behind–the-scenes vexation. We merely say "Accept it, and don't ask questions for which we cannot give any answers." But physics texts are not all there are, and it is at least conceivable that cognate questions, i.e., what is the origin of life and how did highly organized multicellular life occur, which continue to bedevil evolutionary biologists, may provide clues to a working hypothesis.

In the last chapter I discussed the situation where the phote is truly massless and $> >c=C$. Now we are dealing with a situation where $>c$ must possess a degree of mass and therefore cannot equal c.

But what of gravity here? And how does that enter into the already murky discussion of "C" and $>c$?

Shift gears and regard the conventional, distinctly mystical, one where the presence of mass somehow distorts space itself so the mass falls into it, much as ancient mythology claimed that oceanic whirlpools engulfed hapless ships and dragged them to their doom in the depths of the ocean.

Note how this description is essentially two-dimensional and then realize that the statement requires a three-dimensional depiction where the mass is englobed within a sphere of space rather than resting atop a flat rubber sheet of space and thus cannot fall in any direction! As described, the massive object is as fixed in space as an antimatter particle is when confined within a magnetic bottle. The truth is, our failure to recognize the flaw in the flat sheet analogy is caused by our own paucity of imagination; but then flat-sheeting is not necessarily the only explanation.

But what other explanation is available?

Why not try thinking outside the box for a change?

We have mass, and we have gravity. We have excellent operational values and descriptions for both of them. But we have virtually no theoretical explanations for either; at least not apart from those which would make them notional, or perhaps escapees from some other dimension.

But mass produces gravity and gravity acts on mass. It sounds like a 'which came first, the chicken or the egg?' sort of problem. So why not try combining the two? Start with the premise that no phenomenon can interact with another if there is no point of congruency, ergo, mass and gravity must have characteristics in common. Try thinking of matter as 'frozen' gravity, or of gravity as a simple manifestation of mass, and see where that gets us.

This idea is not as alien as it sounds. We think of steam and ice as possessing a common nature....with water standing somewhere in the middle. Nothing alien about that. So why not regard mass as merely frozen gravity?

Would it totally disorganize the physics racket? Try it on for size and see what we may find.

Make a Euclidian type theorem of my premise: No object can affect another object unless there is at least one point of reciprocal congruency with the other... a point/counterpoint motif, if you will.

Codicil: All matter is a product of certain configurations of light at exceedingly minuscule, compressed wavelengths, ergo both mass and gravity are attributes of light.

Conclusion, mass and gravity are both implicit in light and matter is simply one attribute of mass and gravity. This conclusion is already present in modern physics, but the organizational structure and semantics of our science language effectively obscures them.

For the time being I invite you to regard this as no more than a working hypothesis. Later I will have occasion to treat it more seriously, but at the moment I am concerned with the implications of a situation where light has mass and gravity and >c therefore does not equal c. Start thinking of gravity as essentially identical with mass and suddenly the esoteric interpretation vanishes and becomes no more than a latter day phlogiston.

Regardless of the pretensions of obscurationists most physics are remarkably simple. It is just that by the time they have talked all around the subject and then smothered it with arcane symbols meant to awe the plebes there is not much space left of the original ideas. The concepts are still there, but they have been safely buried in a remote section of dictionary graveyards where only the high poobahs of pomp and artifice can recover them for further use. Despite this rather snide remark I must concede that there are numerous occasions where confusion is not deliberate but is nonetheless inevitable.

If humankind was to develop any sort of a science it had to conform to the environment *and* develop a set of universal criteria. And the demons of common criteria persist in bedeviling us poor humans! As a recent example, NASA lost a perfectly functioning Mars probe during its descent onto the Martian surface. Upon examination of the program it was discovered that one set of technicians had been using metric values while a different set of technicians was using the largely superseded 'Anglo/American' miles and feet!

Oops!! Scratch 3 billion dollars and several years' worth of labor, all caused by a simple failure to communicate!

For a different example, it took thousands of years before science developed to a point where the concept of air pressure was clearly articulated; perhaps when soldiers and merchants passing through mountains realized that water boiled quicker but was less hot at higher elevations.

So what is the boiling point of water? It only boils at 100^0c at sea level, and then only when the barometer stands around 29.92. The rest of the time it is something else. Water is a solid, a liquid or a gas, depending upon temperature and pressure. In the heart of a star iron is a multiply ionized gas, etc. Apply enough pressure to ordinary water ice and the ice crystals reform and ice becomes more dense than the water in which it used to float, which in turn means it is deucedly colder than regular ice and it also sinks to the bottom in your glass of ice tea! In a near vacuum the boiling point of water is effectively zero. Etc. There is no consistency anywhere, yet our thought patterns have remained fixed... even among persons in the science profession. Save for a very few of us, we persist in thinking of water as a liquid which occasionally does odd things, such as converting itself into snow, or ice, or vapor, or whatever.

Eventually it became customary to take it for granted that any experiments or statements carried the implied codicil that it is restricted to normal temperatures and pressures, or NTP (i.e. air pressure at or near sea level and temperatures

around 40⁰C). (Note that this avoids the current habit of alluding to it as 'Standard Temperature and Pressure', or STP, as is sometimes done, apparently in the expectation of driving the automobile fuel additive out of business.) But if special temperatures or pressures are involved those must be specified in advance. I will be adhering to these dicta here. Of course, I have no idea what ought to be regarded as normal temperature and pressure in the heart of an accelerator! But *ce la vie.*

We have another source of confusion created by the employment of knowingly false measurements in many parts of science. Astronomy, geography, history and archaeology are no doubt the worst offenders in this respect; but it is essential if total confusion is to be averted down the line.

Political science, of course, is immune to these standards, but it almost never verges on accuracy so my restrictions cannot by any stretch of the imagination be taken as applying to them. My comments on scientific techniques exclude political science and should be taken as such.

Suppose a surveyor seeks to resurvey a land grant initially made when Mexico ruled in Texas. All the figures are expressed in Spanish leagues, while a later subdivision (which overlapped onto an area never surveyed) was measured in French metrics and an even later American survey, which lapped over into the Spanish and French tracts, used the conventional English measurements.   Among latter-day archaeologists try to imagine the confusion when the Moslem calendar, the Christian calendar, the Buddhist calendar, the Mayan calendar and the Jewish calendar get mixed in associated records. An even more puzzling case harks back to around b.c.e. 2,800, following the reign of the Pharaoh Pepi II, which Egypt disintegrated into dozens of rival states. On the walls of the tomb of a scribe there he was reciting the events of his life he prefaces the stark remark "As for Pharaoh, who was Pharaoh? Who was *not* Pharaoh!" Each of the competing states dated it's calendar from the accession of the local pretender!

It is as if some historian might seek to recount the history of the United States if each state maintained its own calendar, which detailed an event as occurring on the 'fourth year of the accession of Lord Huey!' How to correlate this with events occurring in the other 49 states?

I suppose it could be done, but the logical solution is to pick an arbitrary starting point and recalculate all dates around it.

In astronomy, where dozens of people were trying to determine the speed of light or the mean distance of Earth from the sun, where a convenient nearby number is selected and all data is recalculated in its terms, so we have 'star dates', with apologies to the TV series (which co-opted it from a long-standing habit of the astronomical fraternity.

This sort of confusion may perhaps be better understood by a wholly irrelevant but oddly pertinent example. After the Viet Nam War large numbers of Cambodians, Vietnamese and Laotians were brought to this country. In those countries the family name always came first, with the given name last. In the United States it is customary to place the family name last (except for phone books, the military, in filing systems, etc., when we use the logical system of last name first.)

But the American field agents sent over to process these refugees were not uniformly aware of this... or the implications were not understood. Thus a Riem Heng being processed by one field agent might be listed on the preliminary documentation as Heng Riem while a different field agent might call him Riem Heng on a different set of documents!

When he arrived in the Philippines the agent there might assume the agent in Thailand had failed to make the change so Riem Heng once more became Heng Riem. Then when he got to the States and was given his final processing and Green Card his name was again reversed on the record!

If this sort of absurdity could befoul a relatively simple procedure such as the processing of refugees try to visualize the bedevilment it might create in science. For an example direct from astrophysics, the value of the gravitational constant works out to 6.67 x 10^{-8} dynes per cc or thereabouts, but many values have been employed over the past century or so; always using the latest experimental data, and next year's determination may be still different. What is a researcher to do when he is scanning old documents, which may employ a half dozen different values? There is not much he can do when some lineal measures are expressed in "miles", "leagues", "lei's, or "verst's or "stadia", Sumerian cuneiform, Roman numerals or any other system of measuring. This has caused plenty of confusion in the past. The hope now is to prevent it in the future. Of course, us humans being what we are, we won't succeed, but perhaps we can minimize it.

How? Simply by adopting a standard measurement and defining it so all future measurements are based on that. Then researchers thousands of years later can look at that number and crank in the appropriate correction as needed. It won't correct work done prior to the adoption of the NTP standard, but it is a start.

Digressing from our digression, and in the process no doubt adding to the exasperation of professionals at obscuration, there are those who are mystified by the arcane mysteries of kinetics, so a brief exposition is in order.

Call it the intersection of kinetic energy and brick walls.

You have just taken your new vehicle out for a short drive, and being averse to donating unnecessary dollars to the oil chaps and taxing agencies, you are taking things easy. You are cruising along at 60 km/hr when you belatedly notice a brick wall bestriding the road ahead! Now this is not nice. In fact, it is very un-nice, and the result is Galileo's $1/2mv^2$ littering your auto all over the highway.

Nothing daunted, you purchase a new car and try a second time. To your disgust, here comes that wall again! Worse yet, this time it is ambling toward you at 5 km/hr! So now the combined impact speed is 65 km/hr.

As Galileo was so kind as to point out, the impact can be quantified as $1/2mv^2$. But at 60 km/hr the operational number for $v^2=3600$ for the initial collision where V^2 is 4225 while at 65 km/hr smash-up. Rather than being merely .17 percent greater in force it is 1.85 times worse!

But you are nothing if not determined so you grab another car and set out again. You are going to drive down that road if it kills you! Taking pity on your stubbornness and the abuse I've inflicted on you, when you next attempt to drive down that demented highway you manage to sneak up on the brick wall and streak past it before it realizes what you have done. Now it has no choice; it has to chase you! Since its top speed is 65 km/hr it takes a while to catch up, but when it does it slams into you from behind. Your 60 km/hr cancels 60 km/hr of the wall's velocity so the kinetic energy of the impact is based on a speed of 5 km/hr squared, or 25... scarcely six-tenths of one percent of the initial hit!

Summing, three different impacts, occurring under only slightly differing conditions, can lead to enormously disproportionate energy costs, i.e. 3000, 4225 and 25!

Next we perform an interesting little thought experiment and see where it takes us.

Postulate a dimensionless point in a placid pond. Immediately atop this point we place a pair of sensors, each monitoring the surface of the water with their backs to one another. Now we drop a pebble from the point and measure the spacing of the resultant ripples. Absent any outside factors a sketch of the ripple pattern would look somewhat like this:

$$(((((+)))))$$.

FIG 1

The outgoing waves in either direction would be identically spaced. Now let the point be put in motion above the surface and see what the sensors have to say:

$$)))))))))) +)))))))))) \rightarrow .$$

FIG. 2

Nothing has been changed insofar as the pond is concerned, but the sensor is able only to discern that there is a motion in a given direction but, assuming that it knows it's elevation above the clouds hence its speed with respect to the pond can be inferred but it cannot pinpoint the relationship. For examples, you might be heading toward point X at 20 km/hr but it is moving away from you at 40 km/hr so the stupid sensor reports that it is speeding at 20 km/hr. But the reverse might actually be true and you are speeding 40 km/hr while it is chasing you. Or, lastly, either you or it might be standing still while the other is doing all the moving.

Next we consider a third alternative. Here the observing pedestal is observing a running stream. What do we see this time?

FIG. 3

There is no difference between Figures 2 and 3! Absent some outside reference point it is impossible to determine which is in motion. It is even possible to claim that both are in motion and what is seen is the product of the combined motions.

Now postulate that each wave crest exerts one dyne of force. The aggregate of force indicated by the sketch used here therefore works out to 20 dynes. But the compression to the right side of the sketch means that several times as much force will be exerted per unit of time on the compressed side. On the opposite side the force per unit of time will be proportionately less

And that, my friends, is the Doppler Effect. The shorter the wavelength the higher the pitch of noise, the higher the energy content and/or the greater the speed of the system. Translated into colors in a spectroscope, an object which is blue shifted is moving toward you while one which is red shifted is moving away. A blue light contains more energy per unit of time than a comparable red one, etc.

But please note that the energy per phote of this light has never changed. All that has happened is that on one side of the equation twice as many photes pass into the detector per given unit of time while the other side of the equation sees only half as many! There is no difference in the energy per phote; only a difference in the quantity of energy units per unit of time. So now all the pieces are together... which is essentially basic Einsteinian relativity when rendered into English and freed of jargon.

Now it is time to assemble the arguments and ascertain (if possible) to which c-set light belongs, and on that note I tardily return to my cosmology.

Take a mundane little item called the "Poynting-Robertson effect" for instance. As expected, it is named after a pair of physicists who first developed the concept. It is essentially straightforward and easily understood, though I recall hearing a PhD in astronomy ducking a question involving the effect by dismissing it as "a product of Einstein's Relativity Theory"; as if that made it so arcane that the common peasantry could not possibly comprehend it. But the Poynting-Robertson effect is quite important to our examination of light so I shall damn the torpedoes and plough ahead, pretending I know what I am talking about, despite the fact that I regard myself as being a member of the common peasantry and thus an object of contempt by the elite astronomical fraternity.

Suppose I start with a relatively simple chain of logic (which is only fitting since this is supposed to be a relativistic phenomenon). In technical jargon the term "Poynting-Robertson", if reduced to its ultimate source, alludes to: the omnidirectional reradiation of monodirectionally received energy.

So there! You see, I can toy around with exotic lingo when I choose to do so! It is all just a matter of sounding impressive enough to cow any opposition and elicit gasps of awe at the brilliance of the speaker.

Now we translate a bit into ordinary English and see what, if anything, we can glean from this.

When put into plain language all the fancy lingo really says that if you put a tea kettle atop the heat element of a hot plate and turn up the heat, the heat comes up monodirectionally from the bottom and in due course the sides and top of the teakettle get hot with heat leaking out in all directions (omnidirectionally). The Poynting-Robertson effect is that simple! Heat energy comes in from the bottom but comes out in all directions. All the high-flown rhetoric is little more than a snobbish routine aimed at flummoxing the illiterate and impressing them with the sublime magnitude of their professional intellects.

Such is the root of the Poynting-Robertson effect, but we can probe a little deeper and perhaps reduce the matter to even simpler terms. Implicit in this is a determined effort to avoid confronting reality head-on by employing an elaborate circumlocution.

This "radiation" we are talking about is merely ordinary, everyday light in the form of heat. In brief the whole Poynting-Robertson effect may be simply expressed as: "when light strikes an object it shoves it around a bit; and the 'shoving' is merely the release of light at a rather different wavelength."

What is so wicked about saying that? Well, it implies that light possesses a property called kinetic energy.... which merely means it pushes... which means it must contain a trace of something we choose to call "mass".... which means that it must also contain something we elect to call "gravity". And if that is the case the speed of light must be somewhat less than infinite. Thus >c cannot equal C. Now let us expand on things a bit and see where it leads us. I promise it will come as something of a surprise; assuming you are not already neck deep in astrophysics.

Light emitted from source striking object tends to push it away as 'light pressure', but a portion of these rays are absorbed as heat energy. However, there is a limit as to how much heat can be absorbed without the target melting and evaporating. Once this limit of heat is absorbed any added heat must be emitted in all directions virtually as soon as it is received. That which flows out from top and bottom is mostly in balance though strictly speaking there is a bit more push away then push back. When all the received energy is calculated we discover a slight discrepancy. In part it is a product of size. A planet such as Earth may be pushed away from the sun 1 or 2 cm per million years. But the countering force would slice this by 89 to 90 percent so the actual shove away from the sun might amount to around a centimeter per million years.

By contrast dusr motes and gasses may be expelled from the solar system in a century or two.

In strict obedience to Newton's "equal and opposite" dictum, the energy emitted opposite the orbital track provides a slight added boost to speed the over-all velocity of the target so it normally seeks to bleed off the excess speed by rising to a higher orbit. But this is balanced by the fall toward the central emitting mass in order to gain the added speed needed to stay in orbit, so this is pretty much a wash.

Poynting-Robertson enters the picture as a 'fore and aft' effect along the orbital track. Here there is a continuing orbital decay which can only become intensified with time as the target object spirals into the source.

Visualize a remote source of heat such as our sun. Now place a sphere, one meter in diameter to represent a one meter globe in orbit about the sun at a distance of 150,000,000 km (Earth's rough orbital distance). The globe is simply a balloon filled with water, which tells us it has a mean density of 1.0.

If the water bag is abruptly inserted into a stationary position with respect to the sun and locked into that position, then the water will warm until it reaches a point of radiative equilibrium where it is radiating heat out as fast as it is receiving it from the sun. This cannot be a source of argument simply because you cannot keep pumping energy into an object forever. Try doing that and you discover it ultimately just gets hotter and hotter until it reaches either its maximum capacity for retaining heat or, alternatively, the same heat as the sun... or, for a third alternative, the steam pressure from the boiling water bursts the bag and it dissipates as a thin mist in the vacuum, losing a lot of heat in the process; just as on a hot day the air inside the tire of a fast moving vehicle can overheat and burst out. But if you depress the valve stem and let the air escape it comes out colder than it started.

Same principle.

Even solid tungsten steel faces the same limitation. Get it hot enough and it reacts by melting or vaporizing, in the process cooling somewhat. Very simple, very straightforward. Anyone who wishes to may crank the arithmetic into a laptop computer and call himself an astrophysicist, or whatever. It is what happens next which makes Poynting-Robertson important; and different.

Returning to our water bag, we next give the sphere a solid nudge so it is in a circular orbit around the sun at the 150,000,000 km starting distance... which works out to an Earth-range circular orbit at a speed of around 30 km/sec.

What happens?

The easy solution says 'nothing happens.' It simply circles around the sun for ever and ever. But experience teaches us that in real life the simplest solution is almost invariably the wrong solution, and this dictum holds true in physics as surely as it does in politics and economics. You can pretty well take it for granted that any glib, easy solution is going to turn around and bite you... which is why politicians (including scientific politicians) keep coming up with them as sucker-bait to confuse and beguile the voters!

The energy being radiated from the globe precisely mirrors the spacing of ripples around a stone dropped into water, viz:

(((((((((((x))))))))))))

FIG. 4

The heat (light) flowing out from the leading face of the orbiting bag is blue-shifted to reflect its 30 km/sec speed while the energy being emitted from the trailing face is red-shifted by a like amount. The two are not in balance and the speed around the sun is slowed because the leading face of the globe is subjected to a greater equal and opposite retardation than the trailing side. The result is inevitable. The globule slowly spirals into the sun. Why? Because it is now going too slow to sustain the orbit it is in. It therefore must regain enough speed to meet the requirements of the new orbit it is moving into. But this unbalance persists no matter what the orbital distance may be, which in turns means a slow, decaying spiral to extinction which must continue accelerating with each meter of orbital reduction.   I do not vouch for the accuracy of the next statement here, but I recall around a half century ago reading where some physicist calculated that the one meter sphere with a density of water placed in Earth orbit around the sun would spiral into the sun in about eight million years. So there is no reason for anyone to lose sleep fretting over the matter since none of us will still be around to watch it happen, especially when we consider that our old Earth is immeasurably more massive than a mere water bag. It is doubtful whether the Poynting-Robertson effect has prompted Earth's spiral into the sun to shift us more than a dozen meters nearer the sun over the past 4.5 billion years, so everyone can rest easy on that score.

But there is actually a bit more to be said here. Back after the Korean War but before the space program began a pair of Soviet physicists, jointly named Yarkovsky and Radziewsky (don't hold me accountable for the spelling since I have never heard more about them) pointed out an exception to the Poynting-Robertson analysis, viz: it only works if the water-bag is in synchronous orbit about the sun so it always keeps the same face pointing to old sol.

Now to figure out what happens when it is also spinning on its axis?

Ever watch a fireworks pinwheel? It is exactly the same thing. The only difference is extreme slow motion at the start. But when it gets really wound up it is pretty well whizzing along.

Reacting to the centrifugal force our balloon-bag gradually flattens out until it resembles a pancake with the perimeter aligned edge-on to face the sun. Hang on a bit longer and the angular stress exceeds the structural strength of the bag and it disintegrates.

So much for Mssrs Yarkovsky and Radziewsky. But if we like we can extend this line of reasoning by asking ourselves what happens if the globule is rotating with its spin axis pointing directly at the sun? I have never really worked it out in my mind, but it should be extremely nasty work, depending on the rotation, or lack of it. Since it is irrelevant to the present discussion I shall thankfully decline the opportunity.

Now it is about time to revisit the question of the speed of light as possibly opposed to a Cantorian infinite speed. There are a number of options, all of which are mutually exclusive, so we might as well go about it systematically and see what happens.   The number one possibility is the conventional physics premised more or less on Einstein's model. In this we have an imaginary gravity created by some sort of warping of space. Light has an imaginary mass which allows radiation pressure, the Poynting-Robertson effect, and perhaps a few other activities I have not dwelt upon here. Mass itself is undefined but its effects are known and there is some evidence deriving from accelerators which suggests that it resides in a specific cluster of bosons, though how it manages this remains a mystery. Conceivably it is some sort of a parasite with a taste for bosons, or perhaps it is merely leasing advertising space on them.

This approach carries a considerable load of baggage along with it, perhaps most importantly, a logical necessity to posit a fabric of space where every 'particle' of space is connected to every adjacent particle in a vast, spider-web of invisible, insubstantial somethingness. Additional is a need to talk in terms of 'imaginary' mass an insubstantial gravity, along with a wholly mysterious e.s.p. sort of continuing connection remaining after a beam of light has been split apart.

This last is a consequence of recent experiments where a beam of light is split, with half being sent along a parallel track but tending to react appropriately when the other track is manipulated. I am unaware of the precise separation between the two semi-beams but presume it was not appreciable… being perhaps measured in millimeters rather than meters or more. As a footnote to my doubts, I am equally unaware of whether it was the research team or an eager headline writer who alluded to the findings as a sort of e.s.p.

I confess my uneasiness at this. To me it verges uncomfortably close to a return to astrology rather than cosmology. But it is definitely an option, and the new mysticism may foretoken a return to a figurative burning of scholastic heretics who fail to conform to the party line... a modern return to Lysenkoism if you like. But I regard this last as unlikely. More likely is a sort of academic purdah where doubters are tut-tutted and thenceforward ignored for their heresy. Still, it may be correct in part. Most probably, the data are correct enough, but the way we interpret that data may be seriously flawed rather as an earlier generation of astronomers erected a magnificent edifice to support the Ptolemaic universe.

To repeat, that is one option; the one which currently occupies center stage in the physics game. But there is a second suggestion, which is not altogether original with me; and one which I favor. It is based on the premise that the light phote possesses a minuscule amount of mass and thus of gravity.

I know I discussed this a few pages ago, so there is no point in rehashing it. There are, however, a few arguments against this approach, and those arguments ought not to be slighted. Most significantly, repeated efforts to detect mass and/or gravity in light have come up empty. This would seem to rule out the alternative and therefore leave us with no viable choice save to accept the first alternative; even if it is rather clunky and aesthetically unsatisfying in its need for some tricky assumptions.

But there is reason to question the decisiveness of the argument against this alternative, and there are some legitimate questions here. These we must address next.

For one, our tools may not be up to the task when it comes to either extreme of the electromagnetic spectrum and it may well be that the solutions to these problems are currently beyond our reach. This is a technical issue which may be overcome by the ingenuity of scientific engineers and as such, it may be more of a rhetorical objection than a substantial one. But it is only fair to point out that we have no tools to probe the characteristics of individual photes and may never be able to develop them.

A more subtle problem has not to my knowledge been addressed; nor even contemplated. In essence, we focus all our attention on the wavelengths of light and never ask what causes light to have wave-lengths.

Yes, we have excellent knowledge of absorption and emission spectra and some understanding of the spectrum as seen through a prism. But no one seems to have any idea how a prism shows the same beam of light to be simultaneously advancing and retreating

I address this next. It is more complex, but it is certainly an interesting idea; one which may be correct despite the seemingly appalling unorthodoxy when first viewed.

VII

THE DARK OF THE LIGHT

There is even a subtle clue underlying the matter which has never been noted, i.e., the absorption lines in a spectrum. This may sound like a ridiculous comment in view of the attention being paid to absorption and emission lines from stars and the elaborate equipment being employed to measure these lines. But suppose we pause and consider the matter a bit further. These and related questions comprise the present chapter.

As a preliminary comment, I know that my remarks here will be less than meticulous as well as mainly philosophical, but to the best of my knowledge I am treading on terra incognito where no research has ever been done and few questions have been asked. This leaves me with scant meat to chew on so I have to rely on logic. Sorry about that.

Visualize a prism with a light shining through it. Now place a piece of paper before it, In due course a beautiful little spectrum is all laid out before our eyes. In this spectrum are absorption lines, created when gaseous elements sitting in the way of the light beam appear to absorb and thus cancel out fragments of that beam, leaving black voids in their place. These void lines are as specific to an element and its isotopes and their excitement levels as fingerprints are to a man, so once we have identified the lines we know how to recreate them.

When we excite the same element it by heating it to incandescence it emits light in the same pattern but brightens the same beam of light. In effect, we can take the same beam of light and subtract little sections of it or add a little to the same region. We even have terms to define the phenomena; calling them absorption or emission spectra. Now suppose we ask yet another question; i.e., is it possible than the beam of light has been chopped in two when an absorption line turns dark?

I am aware that bright and dark are enhanced by adjoining contrasts. Sunspots, for example are not intrinsically dark and are in fact quite bright…only their surroundings are so much brighter that they seem dark by comparison. By analogy, perhaps the same applies to the dark lines on a spectrograph. Possibly they are still bright but are overshadowed by the even greater brightness of their adjacent bands.

But consider this. If the line turns brighter when an excited element strikes it we have obviously added something to the spectrum. It therefore seems equally obvious that the dark emission line denotes a place where something has been subtracted. This implies complexity to the light, which is not something emphasized in the textbooks… or at least I have not encountered any which address the structure of light.

Despite this subtraction the light beam as a whole continues to remain coherent, or at least seems to. Add to this the proposed e.s.p. of a beam of light and it becomes obvious there is more to be learned here… with no theory already in place to point the way.

So what occupies the space of the dark line?

Is it possible that we have here a minuscule fragment of the mysterious "dark matter" which currently bedevils astrophysicists?

Perhaps. But perhaps not.

Which brings me to yet another digression?

Bearing in mind the possibility that our tools may not be as discriminating as we might wish, and adding the prospect that we have focused all our attention on the wave patterns of light to the virtual exclusion of all else, I now temporarily exit the problem of wave spectrums and concentrate on the light itself. This changed focus is analogous to focusing on water *per se* rather than the forces which create waves in water.

Visualize a single phote of light. At this moment we do not attribute any wave-length to it. It simply is.

What qualities must this phote possess? What are its attributes?

Bear in mind the Cantorian limitations here. If this phote is the solitary occupant of an all-enveloping nothingness, then it is simultaneously infinitely small *and* infinitely large. Either way, its ingredients must comprise the totality of our postulated universe.

Not only is this postulate inconvenient but it achieves nothing, apart from adding a needless complication to things by requiring us to posit a new original phote contained within the present one and thus repeat the process, ad infinitum.

Accordingly, I opt to regard this phote as being next to infinitely tiny, with only a minuscule separation between it and nothingness.

Still according to my postulation, this is all there is; 'There simply ain't no mo'. From this solitary phote we may deduce certain essential attributes if we are to arrive at the universe we perceive around us. For one, it must be amorphous and capable of deforming itself, transforming itself between globular and extreme ellipticity, verging on

linearity if called upon to do so. It must also carry a charge and, if multiplied by others of its ilk,
Exhibit the potential of linking arms, as it were, and forming a loose chain.

Necessarily included in every phote is a minuscule potential of mass/gravity. And this necessarily leads to an inescapable group of consequences. We have already postulated that we are focused on the alternative where >c does not equal "C" Now suppose we see how these consequences must play themselves out.

Begin by conceptualizing an indefinitely long series of high-powered searchlights aligned to create a straight line in an otherwise empty universe. Next we introduce a tiny sphere, which we posit, albeit incorrectly since it is moving minutely slower than C, but jogging along at nearly the speed of light, moving from left to right at the bottom. The right hand vertical denotes the light speed barrier.

Figure 5 is admittedly flawed is several respects. But this is deliberate on my part. There are three layers of complexity to consider and it seemed better to address each of them in turn in order to avoid confusion. So here goes nothing.

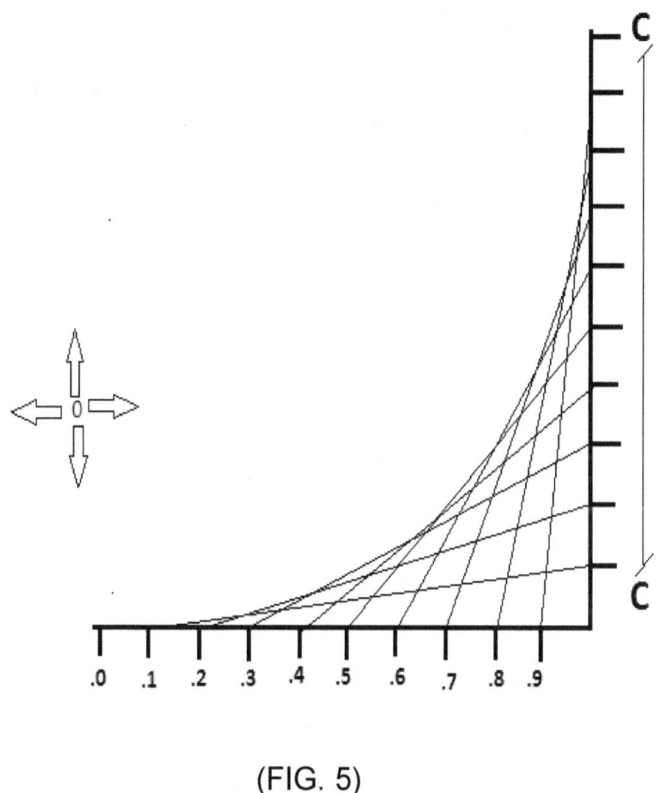

(FIG. 5)

This little sphere is large enough to intercept some of the light from the searchlights, so radiation pressure tends to repel the sphere, pushing it at right angles to its initial trajectory and accelerating it along the path of the searchlights' rays. The process is slow at first, but it is continuous and with every passing second its repelled speed increases. Now we shift gears and imagine ourselves to be astronomers seated atop our little sphere and

121

studying the searchlights. And what do we discover?

Lo and behold, we see the array of searchlights are busy fleeing from us! They are all running away… and the speed of their escape is a function of their distance from each searchlight! The greater the distance the greater the speed with which it seems to be accelerating away from us!

Evidently we have discovered the Hubble Constant!

The further the sphere progresses along this track the more distant the searchlights become… and the fainter in luminosity and the redder the spectral shifting we see.

But this is a flawed version of the constant. It is flawed because relativistic factors enter into the picture. It is merely one of the several factors at work here.

And what might these be?

We have introduced light waves into the picture, and my initial postulation excluded outside phenomena. At the start it was a fairly good idea, but it could be applied only when there was no more than a solitary phote, and now I have increased the complexity. Somewhere in this the rays from the searchlights manufactured wavelengths. For the moment I will let this pass, just to avoid having to make digressions to the digressions from my digressions.

Instead we turn back to our line of searchlights. There is even more information to be extracted here.

Start by postulating that in the beginning the little sphere was aimed straight toward a remote destination so no vector is involved. But at any arbitrary Point 'A' an onboard observer note that his goal has been displaced by 1 degree. A later observation may deviate a 3 degree displacement and the next one after that shows a deviation of 9 degrees!

Lo and behold, this proves that space is curved! If there is any residual doubt, visualize the conclusions of a hypothetical observer situated at the putative destination. The spheroid obviously was aimed at him, yet it is persistently deviating away from its goal at an accelerating rate!

There can be only one explanation: "Space is real. It is not a void. It is tightly woven and governs the trajectory of everything passing through it! " In summation, it is possible to arrive at substantially the same conclusion as current orthodoxy depicts without having to postulate all sorts of esoteric physics.

There are so many diverse factors in the logic it can be burdensome to strew sidebars all over the landscape. Accordingly, the question of the origins of wavelengths can be deferred for a space. This said, suppose we return to the relativistic question.

Go back to our rogue automobile and the kinetic energy generated by its efforts to get past the wall. The smaller the difference in combined speeds the less the kinetic energy profile produced. If our little sphere is accelerated to 296.9 km/sec along the right angle track of the kinetic energy of the searchlight beam will be somewhere around 1.0 km/sec (0.001 percent) of the force it exerted on the spheroid at its point of entry onto the graph! In short, the minuscule force of its radiation pressure at the start will be reduced to a point almost imperceptible, and it can continue on its new vector indefinitely without ever succeeding in reaching C. At most it can accelerate the sphere to its own speed, which we have defined as >c.

Even ancient Euclid gets into the act with the inverse square ratio.; which probably dates back to old Sumeria. Newton refines it, Galileo confirms it, and countless students and professionals have added to the list of independent confirmations. And most importantly, it is logically impeccable, with no fudging the mathematics by slipping flawed "iffy" postulates into the mix without making them explicit.

Conclusion: The faster our sphere goes the less the kinetic energy transferred to it by the pursuing rays from the searchlights and therefore the less the spheroid is accelerated, ultimately being reduced to a crawl where it's speed increase by less than a millimeter per century or so. No matter how many lights are focused on the sphere it can never be accelerated to C. There will always be a minuscule difference between >c and C. It might be an infinitesimally tiny shortfall, but it will be there. It is Zeno's old paradox of Achilles and the tortoise all over again. It makes perfect sense if we go by Cantor's rules. So Cantor's logic remains impeccable.

As the ever-so-banal TV commercial invariably orders, "But wait!" before promising two of the gimcrack items if you order now! I continue by promising, "But wait! You ain't heard nuthin' yet!" In the first version of Fig. 5 the left hand starting point was presented as an object in lineal motion at immediately sub-C speed and subject only to radiation pressure and the Poynting-Robertson effect. And how does this play out in our little scenario? Throw a taste of Doctors Poynting and Robertson into the mix and we discover roughly the same rules in play. At the start of its journey to C the target sphere is nearest the searchlights so it absorbs the maximum amount of energy and thus the effects of the imbalance will be greatest. With increasing distance the imbalance wanes but does not vanish altogether until all further momentum toward C is lost and radiative equilibrium is achieved, at which point the light from the source becomes invisible because our little balloon is receding as fast as the light rays chasing it and thus can never catch up.

Between the two forces of radiation pressure and Poynting-Robertson we can chart a beautifully curved trajectory. Cantor's logic still remains impeccable.

Interestingly, the equation balances this way, where it fails this test with conventional physics. Consider the factors involved. First tis the kinetics of the orbital speed. Then factor in any residual energy absorption capacity of the object. The reradiation of energy must be uniform so the quantity of radiated energy can be calculated. The speed of the radiated energy can be taken for granted since it must always be >c. But the kinetic energy account must equal the energy input.

So now we shift gears. Eliminate the searchlights and simply perch ourselves atop our tiny pebble and proceed to accelerate from a hypothetical 0 speed at the left to C at the right while measuring the speed in terms of the distance remaining before arriving at our goal. For this we have logically and observationally firm data telling us that no matter how fast we may be travelling the speed of light with respect to us is always the same 297,000 km/sec; give or take the most recent adjustment.

Einstein resolved the seeming paradox by proposing that as the spaceship, or whatever, approaches the speed of light it is foreshortened, getting shorter and shorter every step of the way until it is virtually a flat plate plunging toward its goal. Ultimately, its lineal dimension as seen by an outside observer, becomes mere microns in depth! But through all of this the occupants experience no sense of distortion. So where does this leave us?

(NOTE 1: Regard this as conceptual rather than accurate. Visualize a kilometer long spaceship enroute to C. At zero speed it occupies the entire line since it fills every cm of the space as it progresses from the infinity of slow to the infinity of fast the kinetics of motion lead to a shortening as the accumulating energy bunches at the leading verge. At first it is relatively weak, but the nearer it approached C the more pronounced it becomes.)

FIG. 6a

Relativity all over again, with the location of the observer determining what he observes. Actually, we all experience relativity every day and think nothing of it, e.g. imagine looking at an attractive young lady in a skimpy swim suit strolling along a beach toward you; then think of the same young lady walking away from you on the beach.

Two entirely different perceptions. Two observers, one in the bow and the other from the stern, perceive wholly different things... One focuses on the curves at the top while the other focuses on the curves at the bottom. And that is the essence of relativity theory.

Visualize relativity in those terms and it is not especially complicated. It requires a dedicated academic to make it seem awesome and complex, but at least I have made it exotic---as in exotic dancers! I have no argument with the Einstein solution. It works, and the fact that Einstein resolved it in the fashion he did proves anew that Einstein was fully aware of Cantor's work. Between this solution and the employment of 'C' sets there can be no doubt remaining. It is entirely possible that the two of them were acquainted. They are contemporaries, they were German scholars and Cantor was highly regarded at the time, so Einstein would be aware of Cantor, though Cantor might well have been unaware of a youthful Einstein working at a Swiss patent office. But there is an alternative way of addressing the problem; one which may be a trifle less mystical. Refer once more to; Fig. 5.

In the diagram both the horizontal and vertical axes are subdivided into 10 sectors, each representing 30,000 km/sec *en route* to C. Thus from an origin 0 at the left to C on the right we encompass the extremes of the infinity of speed, i.e. from the infinitely slow at absolute 0 to the infinitely fast at C.

If we think of C as a Cantorian continuum then the vertical axis must form a 90^0 angle to the acceleration [or deceleration] base. As the speed increases from 0 to 0.10 C the perceptual horizon slides upward along the C wall, until at the peak of its speed it is roughly parallel to C with an exceedingly small space separating the two.

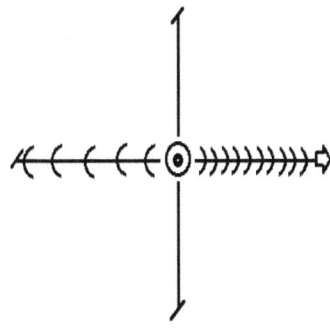

FIG. 6b

This is another case where my sketch is misleading, but now we drop the other shoe and drag Messrs.' Pounting and Robetson back into the picture. Messrs.'. Poynting and Robertson have something to add to our analysis.

Take a closer look at the mix here. Place a star below the bottom of our little diagram and place a minor chunk of matter in roughly earth orbit about it. A portion of the light pressure is absorbed by the sphere in the form of increased momentum away from the light. Another portion is reflected back toward the source, while a third fraction is absorbed as heat, which is then reradiated omnidirectionally from the sphere.

Here again, the reradiation flowing from the trailing face of the sphere will give it a forward thrust. But it necessarily is more than offset by the countervailing retardation of the momentum along the original vector. The end result of this was depicted in Fig. 6b.

Ultimately the near C vector along the original line of searchlights is nullified and the sphere is moving at right angles to the initial axis at a near C velocity equal to the speed of the photes emitted by the searchlights! So here again, we have a reason why >c cannot equal C.

Of course this only applies if we postulate some mass inherent in each phote of light. If we insist that light is massless then it must be tooling along at C, so nothing is proved as yet. We have merely established a few parameters for a mass content in light. So let us see if we cannot add a little spice to the pot.

Begin with the assumption that each phote of light possesses a definable structure, which may be visualized in the form of a tightly compressed coil spring. The energy of compression translates as nascent mass, and thus gravity. So what we have is a tiny cosmic egg of sorts, an egg consisting of energy balled up in a minuscule ovoid vaguely pre-echoing the current model of an atom where the electrons are basically fuzz-balls enveloping a nucleus.

This perception differs from current orthodoxy only in its postulation that the nucleus contains a corkscrew helical structure at its core. But I should add in anticipation that later analysis of the atomic nucleus will lead to a slight modification of phote structure, so what I point out here should be considered merely a minimalist workable outline and not a final statement.

In discussing the journey of a massless phote across the universe we discover that from the standpoint of the phote there can be no sensation of time and the long trek is instantaneous. There cannot be so much as a flicker of internal change in this phote, no blink of an eye, no intervening observations. A camera snapped just as C is achieved but developed on the other end of the trek would be utterly blank transiting from scene A to scene B without any interim pause. It would be a seamless junction which might easily puzzle a hypothetical observer by its disjointedness and lack of continuity.

But none of this would apply if >C possesses mass and thus moves at a speed less than C. Should this be the case then there is at least a potential for a degree of relaxation of the tension holding the coiled energy which forms the phote and thus the photon, which derives from the pattern imposed by the phote. The internal perception of a journey of a hundred billion light years might afford a sense of a mere second, or even a day; a day, during which the wave length of a chain of photes might have relaxed from a deep blue to an intense red.

The phote itself will be insensibly elongated so far as an outside observer is concerned while the onboard observer is aware of no change since the red shifting will have been compensated for its elongation.

An interesting little sidebar here involve the human eye chemistry. To the best of my knowledge no one has ever questioned whether the mechanisms of rod and cone receptors are fine tuned to handle extreme changes in speed as our hypothetical space craft nears >c. Will the crew be blinded amidst a sea of intense UV radiation provoked by the compression of the light?

It may or it may not. No one knows.

This all sounds pretty esoteric, but the interface between infinity and finitude is rather blurred at the fringes and this process merely foreshadows the decay of nuclear particles in accelerators, and even in atoms and may be indicative of the speed of the particles created in our super-colliders. For example, a phote moving at 0.999+ the speed of light might have internal time only for a minor reduction in the direction of total loss of wave length while a neutron might decay in 12 minutes and a neutrino in several billions of years. This is not a new hypothesis by any means. It is merely a renewal of a discredited suggestion that light may become fatigued in the course of its long trek across the universe.

Please note than this does not mean that the light tires because of its long journey. To the contrary, all I am doing is suggesting that many of these particles actually have a common decay time, but this decay time is contingent on the internal time consequent to the time frame imposed by its speed. So 'light fatigue' per se is not so much a consequence of the length of its journey as it is of the ticking of its internal clock.

Rather analogous is the decay of water's wave amplitude as it recedes from its point of origin. The physical speed does not change much, but the height of the wave flattens until it is scarcely more than a ripple after a period of time.

I should also remark that the primary incentive underlying the original rejection of the 'light fatigue' suggestion was not premised so much on any specific arguments opposing it as it was on its inconvenience. In essence, all the current model does is provide a platform to satisfy the prissiness of the bean counters, who can never rest happy until every I is dotted, every T crossed and the books are in perfect balance. But the universe is a fuzzy affair which is utterly indifferent of the arrogance of human bookkeepers and this may be one such occasion.

Add this to the mix and the red shifting of light becomes more complex and less capable of providing a precise yardstick in determining the distance of remote galaxies and the speed of recession etc. So there is really no argument against light fatigue... but neither is there any compelling argument for it. It may or may not be real. But while it introduces a complication by casting doubt on the accuracy of the cosmic distance and Hubble constant scales it offers consolation by the promise of providing a supplemental measuring stick to account for particle decay. The scales may not be wholly balanced, but there is a realistic likelihood that these factors would come near to balancing them.

And this brings us to the ultimate question: Do we have any indication or subtle hint that light might actually undergo wavelength decay, or are we merely indulging in a fantastic dream intended to lead others to endorse our argument?

There is not much out there to support my idea, but neither is there much to refute it. About the only point in favor of light fatigue may be found in the reradiation of light, or in its passage through interstellar clouds in deep space. In both instances the light emerges somewhat reddened. In other words, the wave length may well be mutable, which might support the idea of fatigue. But there is an equally realistic possibility that there are alternative explanations, most particularly momentum transfers caused by collisions with atoms or molecules within the cloud. The trouble here is the likelihood that both alternatives are real and we are faced with a mixture of factors --- which combine to prove nothing!

All of this leads us to a final, and I believe decisive, argument in favor of the thesis that >c does not equal C.

The simple fact is, the current model leads us to a dead end. It solves nothing and has no rationale underlying it. In this respect it is much like the Ptolemaic model of the solar system. Mathematically it works, but it leads nowhere. On the other hand the situation where >c does not equal C leads us directly to the ultimate completion of the electromagnetic spectrum; from start to finish, and in the process resolves problems of magnetism, the missing mass of the universe, and the nature of dark matter which seems to permeate all of space.

Take these assertions one by one. We already know that matter, as a component of ordinary energy, appears in the Compton wave lengths at the extreme left, or shortest wavelengths. Since there is nothing other than light in the universe (save space itself) this matter must be a property of light... at least *in potentia*. Now slide down the electromagnetic scale as it heads toward the longest possible wave-lengths, constantly getting less energetic and fainter until we accept lengths in excess of a few thousand km. Current theorizing alludes to these ultra-long wavelengths as relics of the Big Bang. It may be this, but this is more of an ad hoc conclusion than a truth. After all, simply saying something is true does not make it so --- unless you are a theologian, an astrophysicist or a politician.

There are also a couple of problems here. For example, current orthodoxy firmly rejects the idea of light fatigue yet this same orthodoxy also argues that these ultra long waves are relict of the Big Bang which have flattened out over their long journey through the cosmos! Reconciling the two dogmas probably requires the ad hoc introduction of a mystic L, M, or N force otherwise invisible save for its ability to flatten out the Big Bang radiation.

The other problem is more mechanical. We have 1,000 kilometer light waves, but what comes after that? What lies beyond? Is this the end of the line, or is there more to be found at even longer wavelengths? Is there a point where the wavelength approximates light speed; i.e., where the length between wave crests is "C"?

What would a straight line photon look like? Have we any technique to detect one? And if so, what does it do? Additionally, can the individual photes remain connected once their helically imposed structure disintegrates? Many of these questions will be more completely addressed in Part 2 of my little excursion into those forbidden realms reserved for orthodoxy. But for now I retreat into the sketchiest of outlines. Our instruments are focused on detecting wavelengths, and if there is no wavelength there is no direct detection... though indirect methods for deducing their presence, e.g. gravitational effects, magnetic effects, etc. are available and in use. To illustrate, the iron filing pattern demonstration surrounding a bar magnet and the existence of Earth's Van Allen belt argue that magnetic fields per se may be organized and worked with even though the individual magnetic unit has no wavelength. In fact we do not even know that an individual magnetic unit can be measured. Call this unit a 'mag' and leave it at that.

What we do know is that a string of mags (which may combined to produce electrons) can be aligned head-to-toe to create a magnetic chain, which can then be manipulated to manufacture a wave indistinguishable from any other light on the scale. A flashlight battery takes unwaved energy and converts it into light which is waved, etc. We also discover 'free' magnetic clouds in empty space; clouds with their magnetism intact but apparently disconnected with any source. Nor is this the end of the matter. We have ordinary lightning bolts, St. Elmo's fire (ball lightning), etc. Then there is also all that dark energy occupying supposedly empty space. This dark energy evidently possesses its own mass and therefore gravity in sufficient quantities to affect the entire universe. And if it contains mass and gravity it will possess energy and necessarily must be placed somewhere on the electromagnetic spectrum.

Since the only available slots are very far into the longest ranges of the spectrum we must conclude that this is the residence of dark matter. So now we have a completed electromagnetic spectrum, one which ranges from the ultra-short Compton wavelengths on the left to the ultra-long, effectively straight line lengths on the right.

But watch out!

It is time for another of my "But Firsts." There is a problem in all this and things are not nearly as simple as I have made them out to be. To begin with, there is a little item first pointed out in the 1800's by an astronomer named Olber; hence the title given the conundrum "Olber's Paradox."   And what was his paradox?

136

He merely asked why the night sky is dark!

At the time astronomers still fretted over such items as the great galaxy in Andromeda and believed it to be merely a luminous cloud in our own galaxy. But they did not regard our galaxy as a distinct (and fairly minor) feature in the universe. Instead, they thought of it as an infinite number of stars statistically evenly distributed throughout an infinite universe created by God in order to demonstrate His power and awe his human puppets.

This presented a problem. According to this view no matter where you peer into the heavens your line of sight must sooner or later terminate on the surface of a star, and since this was an 'inescapable truth' then the night sky must be just as luminous as a midday sun and the collective light of tens of billions of stars would engulf our wretched little planet in a sea of heat and radiance as hot and bright as the surface of the sun throughout all eternity.

This was an interesting idea which was debated for several years before finally being laid to rest by two separate factors, plus an erroneous one, i.e., the fact that stars were members of distinct clusters of stars, which are now called 'galaxies', so there was no longer a need to postulate an equal distribution throughout the heavens. Then there was the red-shifting of light caused by the apparent speed of recession from Earth and the consequent lessened luminosity of these galaxies. Lastly was a flawed factor stemming from the large number of dark clouds obscuring huge regions of space?

That objection was dubious in view of the fact that these clouds would also be receiving and reradiating the light from all the stars. But this objection was later in part defeated by the fact that the reradiated light would have lost a significant part of its intensity to the cloud and the cloud itself would be radiating down in the red regions of the spectrum, making the equation balance after all.

So Olber's paradox died a welcome death…. or did it? In part it did. In another part it did not, and in a third part it was transformed into another problem; one which bids fair to being more troublesome to answer.

The part of Olber's paradox which persisted relatively unchanged applies at the cores of galaxies, where tens of millions of stars are crammed into a sphere which may be scarcely 10 light years in diameter with as many as a billion more stars occupying the space out to 50^3 light years!

Lest there be doubt here, I am deliberately being conservative in my estimates. Perhaps not every galaxy houses that many stars at its core, but many most certainly do. And while those stars at the approximate center will not be wholly enwrapped by the light and heat of the surrounding stars, the combined light and heat, together with the thermodynamics of stars, must lead to some very interesting physics and may contribute to the formation, growth and maintenance of black holes in the core regions.

I hope to get back to this problem later in my text but for now I have other fish to fry. It is the third portion of the paradox --- where it is transformed into a new, and more difficult paradox --- which now occupies my attention. Succinctly: I ask why isn't the night sky even darker?

To begin with a simple, and obviously erroneous example, let us consider a single spot on a star where a laser-focused beam of light is sent on an 8 minute journey of 150,000,000 km to strike the moon.

At one tenth that distance even the best laser will have been pretty much dispersed; possibly to a diameter of a kilometer. Given this premise, then an observer standing immediately outside the perimeter of the laser's field of impact will be utterly ignorant of the affair. The night sky will be totally dark. An observer on Earth might see the reflected light from the laser, but he would have no indication that there was more to the moon than what he could see. There are a number of obvious flaws in this argument so don't take me too seriously. But it is a useful starting point for deeper understanding of the internal structure of light.

Individual stars can be seen and photographed at a distance of more than 3 million light years in the Andromeda Galaxy, yet we can see the photons of light coming from those stars. Now we ask how many photes of light must be emitted each second from the source in order to maintain coherency over so long a period of time and so wide a spherical distribution?

Clearly, the number must run well into the trillions, quadrillions or even more per square centimeter of surface of the star! And bear in mind that this is a continuing release every instant. Which leads us to question whether there may be merit to the wave theory of light propagation.

Start with the concept of individual photes and see where this must lead us. Are such numbers even feasible? To answer the question quickly; yes, it is possible. I occasionally take an acidophilus tablet to settle an upset stomach. Each pill is roughly the size of an ordinary vitamin pill, yet each pill is guaranteed to contain over 100 million active Lactobacillus Acidophilus bacteria!

Scaling down, each individual bacterium consists of at least ten dozen molecules of greater or lesser complexity. Each of these molecules probably averages out to 3,000 or so atoms, which in their turn boast in the neighborhood of 100 each of protons, neutrons and electrons.

All in all, it is an impressive tally, one which leads to an absolute minimum of around one billion photons per pill, with a strong likelihood the actual total may be multiplied, or even squared. I deliberately avoid claiming exact calculations here because any figures I use are subject to revision with the next test. What I am mostly concerned with is the underlying concept, and the concept allows for the idea presented here --- barely.

But I do not like it. It seems altogether likely that at even greater distances we would be able to detect images of stars winking on and off as random bursts of light simply miss our telescopes. So what is there available to replace my flawed reasoning here?

Perhaps we have neglected to carry the idea of a wave to its inevitable end. An ocean wave is not a thing in itself; it is a condition where trillions of individual molecules of H2O along with various salts, impurities, etc., are moving synchronously in lock step toward a common destination. Viewed from this perspective the problem is at least mitigated. Additional mitigation stems from our habit of conceiving a beam of light as two dimensional. But the development of the toroidal universe --- the smoke ring --- imprints itself on the photes which go to comprise it. When they align themselves *en chain* they form a corkscrew helix!

And where have we encountered this in our science?

Everywhere! Our very DNA consists of a corkscrew helix. Cyclones are toroidal, as are radio waves (i.e. 'sidebands'). And this merely scratches the surface. So why should we be surprised at the thought of the photes in a beam of light being arrayed as a corkscrew helix? If correct this may help understand the "ESP" of light as well as accounting for the failure of Fraunhofer lines to fractionate the unity of spectra. It would at least be a step in the right direction.

And now to round off the problem of light, shall we address the ultimate question and see where it leads us?

Among other things, Fig. 2 illustrated an alternative to Einstein's spatial compression when matter approaches the speed of light in cases where the speed of light (>c) possesses mass/gravity and is therefore incapable of reaching C. Given this premise, Einstein's light also would have to experience compression as it tools along at a barely sub-C speed. We lack observational indication of this. Of course this does not disprove anything, but on the other hand, the problem never arises in the model I have proposed. This is all well and good, but it raises a new problem.

Visualize the implications. I have suggested a boundless wall preventing speeds in excess of light and extending beyond the torus into the ylem. It would be a wall approached sooner or later by every hint of light in our personal little universe which spans a mere 23 or so billion light years if we assume symmetry between our side of a Big Bang and the other side.

Of course I am setting up a straw man here, a scarecrow to be toppled by the incisiveness of my intellect. Since I am averse to straw men and am uncertain of my intellect, suppose I drop the matter and simply say there are some problems to consider.

It is a puzzlement. But occasionally ideas come from the most unexpected places, and if they pan out a considerate individual will make a point of giving appropriate credit to the source of his inspiration. In this case the source of inspiration stems from a brief furor arising more than 20 years ago, when some mathematician cum physicist (I never heard his name; at least not in that context) proposed the *Tachyon*.

This was supposed to be a particle on the far side of C, but symmetrical to this side. In other words, when an object reached C and crossed beyond it would commence slowing toward an ultimate 0 speed! It was the *yin vs. yang, black vs. white* dichotomy all over again and every atom on this side of C had a counterpart on the other side.

I never saw the underlying mathematics but I assumed then as now that they were formally correct although not conforming to reality; which is not unusual in physics, as witness *phlogiston.* In effect, $1 + 1 = 2$, but it does not say '2 of what'.

Being already thoroughly acquainted with Cantor, I was not impressed. If the author of the hypothesis was correct then Cantor was wrong; and Cantor's logic was too good in too many different contexts to allow me to abandon it on such slim evidence. Accordingly, I dismissed the idea and thought no more on it until quite recently; 2011, in fact. I began putting this text together in 1992 and sought to retrace every step of the logic time and again, working systematically from one conundrum to the next. Where I could find no compelling argument to a problem I confessed as much and went on with the text. But the idea of an infinite projection of a light beam as it paralleled the C barrier simply felt wrong! It was a glib answer to a stubborn problem, and glib answers are almost invariably wrong answers.

I persisted in this unsatisfactory 'solution' for more than two years, then went back to analyzing the problem anew and almost immediately recognized a whopper of an error on my part!

It began with the observation that I was attempting to merge two distinct C sets into one when I put the light beam which I have concluded belongs to the C set of curves into an infinite C set of straight lines, ergo, either the wall is curved or the proposition must be false to fact. But either way the light track must be curved. If the two C sets can be merged then pi cannot be an irrational number!

So what does this mean?

Science fiction writers often point the way to the future, but not always. There are occasions when they are egregiously wrong; thus proving they are human along with the rest of us. The announcement of tachyons came as a welcome thunderbolt, validating years of "Warp ten, Scotty!'" which allowed starships to span the galaxy in a day or two. Tachyons were the key, and the writers got busy.

But there was a problem here. The tachyon was supposed to be a mirror image of the photon on the other side of the C boundary --- which meant it had to slow down rather than speed up --- and that defeated the idea of warp speeds!

No matter this; it was quietly ignored and other means of exceeding C were employed. So scratch that idea. At least a superficial analysis of the tachyon model argues that the instant a tachyon slows to less than C it promptly falls back into our universe. But there is a less obvious possibility calling for a barrier on the *other side of C;* one which prevents a fall-back option.

To retain symmetry the tachyon would have to slow to absolute 0 speed and then translate over to a negative speed, one which would thrust it back into our universe to commence a new acceleration in a gigantic yo-yo game. The universe is literally a bouncy ball where a beam of light is perpetually wavering above and below C!

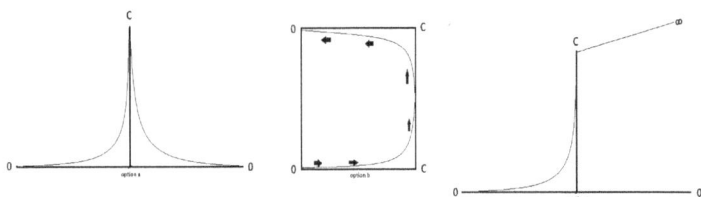

FIG. 7a

NOTE: A spaceship launched at zero speed from the origin at lower left is indifferent to tachyon concerns and remains indifferent to alternatives a, b, and c. but when C is achieved we have three alternative routes, a, b, and c.

We consider the type c 'yo-yo' alternative first. There is precedent for such a concept contained in the idea of a yo-yo universe where a Big Bang blows everything apart and it expands until all the energy is spent and it falls back, only to reform itself and start all over again.

But there are flaws implicit in both of these hypotheses. The yo-yo universe has to be a finite process because the light which passes the boundary of the universe is lost forever in infinity so the energy content of each successive explosion is reduced accordingly. Ultimately, after X number of repetitions, there is no longer enough recovered energy to do more than utter a faint 'Pop!'

In addition to this is the recent 'discovery' that the rate of universal expansion has been increasing rather than decreasing as it would be if the gravitational forces were busily at work slowing things down. This resurrects the earlier model of relativity offered by Einstein in which he spoke of a "cosmic repulsion factor". He later recanted the idea but here it shows up again, all decked out in fresh attire.

Assuming an error in the finding and we actually have a yo-yo universe then the final end to everything. To paraphrase; 'this is how the universe ends; not with a bang but with a whimper!' Turning now to the science fiction idea where acceleration continues infinitely via some sort of space warpage, this too is fatally flawed no matter how we look at it. If we postulate something analogous to an aircraft breaking the sound barrier then Cantor is dismissed and C is no longer infinite. But this violates Newton's dictum by arguing that there is no inherent reason to postulate that for every action there is an equal and opposite reaction. As for the yo-yo tachyon, there is no hypothetical basis for sub-zero degrees Kelvin, negative speeds or any other limits in any direction.

This would seem to preclude all possibility that the tachyon argument has any validity despite anything the mathematics might say. Now comes another, "But wait!" All three options seem to be foreclosed. But as the advertising huckster or hack politician utters his hoary mantra, perhaps I am being a bit premature in my rejection.

Take a closer look at my reasoning and note that in studying the tachyon I am compelled to employ the terms of lineal infinities to examine concepts unique to what we have an already determined to be a curved universe.

Given this it should no surprise to learn that we have come up with nonsensical answers!

There is no theoretical reason for photes in the background ylem to be constrained to obey the rules of the smoke ring. To believe otherwise would be tantamount to arguing that the air through which a smoke ring is moving will be coopted into joining the ring. Lacking any data of conditions within the ylem other than a requirement for disorganized turbulence we should concede that it is at least possible that the underlying postulation of a "warp" drive may exist if suitable vehicles were encapsulated within magnetic fields which transformed them into 'mini-universes' gliding through the ylem. This is the stuff of science fiction, but it is at least a sop to the imagination. It is not likely but we would be exceeding our limitations to deny the possibility.

Just as I was about to give it up a new thought snuck in: Granting that the mathematics for tachyons might be correct is it possible they may have been misinterpreted?

This was the needed breakthrough. But it needs some explanation or it can mislead us.

Perhaps the mathematical symmetry is *not* to the far side of C but instead is on this side! What then? What does this lead to and how do we depict it? Is there some workable solution I am overlooking? Figure 7b illustrates my meaning. The universe is three dimensional and curved so what goes around comes around! But there is more to the interface between C sets than I have suggested. As already mentioned the case where the same straight line may serve as both pi and the hypotenuse of a triangle, it is also possible to inscribe a square within a circle or a circle within a square without violating the protocols of transfinite sets.

Now suppose we look to see if the conceptual rules for 'tachyons' might look from this perspective. Start by taking a fresh look at some of our earlier diagrams and studying them item by item before recombining them. Perhaps we can learn something that way. We can never be certain until we see for ourselves... and even then we must confess that some new discovery a year or so from now may relegate everything we say now to the trash heap along with phlogiston and a flat Earth floating on the back of a turtle in an infinite ocean; both of which were hailed as the epitome of science in earlier eras.

Figure 7b is our jumping off point. Now we must hope for a soft landing

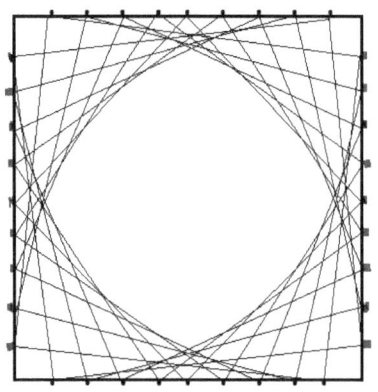

FIG. 7b

Here we have a moderately inaccurate depiction of the internal optical paradox confronting a physics professor on Earth as we orbit a sun which is itself in orbit around the center of the galaxy which in turn has some unknown trajectory within a universe of unknown size and motion as it drifts through an ocean of ylem. The vertical line at the right is the C barrier. The horizontal line at the base is also a 'C' line and the sum of all the Earth's motions (which combine all the motions mentioned above plus any others we may not have mentioned}.

You may postulate any speed and direction you wish, but here I have opted for a starting velocity of 0.1C and have increased the speed by tenths to C to depict the perceptual path slides up along the vertical 'C' line accordingly.

Even in the laboratory measurements of the speed of light are invariant regardless of the speed of the galaxy, its rotation, the direction of Earth's orbit about the sun or any other factor.

This is a paradox for us to wonder at. Study the diagram any way you wish. Begin with an object with zero speed at the onset and accelerating .toward C. The diagram remains accurate. It is reality, not a mere illusion. So how can this be? Start with a beam of light while I take measurements in every direction. Nothing changes. Direction matters not at all. Up, down, fore or aft, from one side or to the other, we get the same answer.

The solution is obvious. We are inescapably caught up in Cantor's transfinite universe. Now drop the other shoe and consider the matter from the standpoint of an observer perched atop a phote.

This is not to say there are no further problems here, and to explain it I must confess a small cheat on my part. When boiled down to its simplest terms the model simply doesn't work!

Recall that I have repeatedly stressed that employing one geometrical C set to explain a different set inevitably leads to error. The diagrams I used were exclusively lineal but the light we were employing was moving along a curved trajectory. While I concede that this is not yet proved the indications developed along the way have uniformly argued for curvature and the straight line model depicted in figure 7c points to an inscribed circulate infinity.

Stripped down   figure 7e emphasizes the resemblance to the hole in a doughnut or the 'eye' of a smoke ring and suggest the need for a transformation of coordinates not unlike the transformation required for altering the Ptolemaic understanding of the universe to be replaced by the Copernican. Place the origin at the center of the doughnut hole and Calculate from there. Inscribe the circle and make note how we are back to the conundrum propounded by Pythagoras.

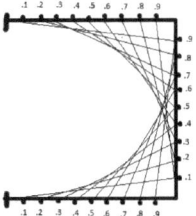

F1G. 7c

Here the postulation accepts the basic premise that the speed of light is the same in all directions. As such it may be regarded as an all-encompassing boundary. Now we make a single stipulation. Since the speed of light is always invariant then no matter at what point on the perimeter of the box an observer is standing his line of sight for measuring the speed of light must remain unchanged, Figure 7c illustrates the results in a reasonably cogent manner.

Remaining in our transfinite reckoning posture we may also note that no matter where the observer may position himself he is always finds he is stationed on a perimeter and thus C remains unchanged everywhere.

About this time venerable old Cantor kicked in the door to explain. Infinity is necessarily a two way street. Infinite speed (C) is outside the time continuum. But infinitely slow speed is also timeless. Infinitely large is the same as infinitely small. Infinity of numbers necessarily stretches in both directions, etc. So if this is correct then we discover that our universe is sandwiched in between two infinities, both of which are characterized by an infinite speed projection ranging from 0 to C, infinite time as a separate C set but the same on either side of the wall; and infinite space throughout!

But the idea of our finite, toroidal, smoke ring, universe being sandwiched between infinities does not make a great deal of sense, largely because there seems to be no inherent reason for the light beam to recurve back upon itself. This prompts us to look into the situation a little more deeply by expanding our horizons.

The sketch on Fig. 7d bears a striking resemblance to that presented by Fig. 7c. We have a 0 origin to the left, acceleration line to the right with a C across the top, but nowhere it there any suggestion of a reason why the light curves back. Does this mean light has a passing fancy and merely feels like making a two-way journey or is there some sort of a mystical mirror effect in play which causes it to reflect the light back upon itself? To answer this we look at Fig. 7e to complete the picture.

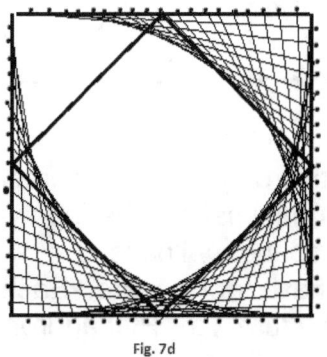

Fig. 7d

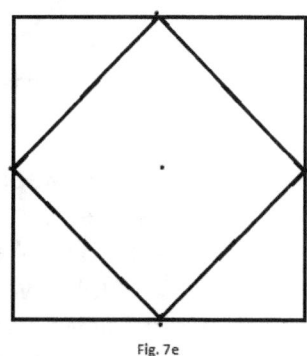

Fig. 7e

FG. 7d & 7e

The solution is obvious. Our universe is contained within the same infinities on all sides so regardless of where we set our sights must appear to be traveling at the same speed so the fault is not with the universe; it is with our perceptual processes *vis-a-vis* infinity.

It involves the seeming paradox where we extract portions of an infinity and use these extractions in a decidedly non-infinite fashion, e.g. we can say that a yardstick is 36 inches long despite the fact that '36' is meaningless when cited as merely an element in the infinity of numbers.

But the picture is still not complete. As it has been presented here it is misleading and our process of breaking it down requires one last step before we can reassemble it into a coherent tapestry. A figure 7e is needed.

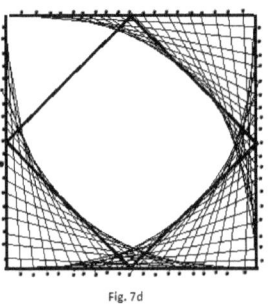 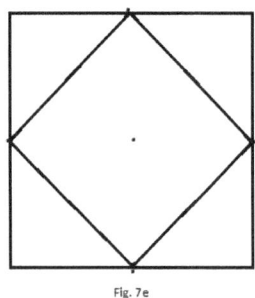

Fig. 7d   Fig. 7e

FIG. 7e

Still keeping the origin at the center of our box and employing a decimal numbering system so the midpoint of each boundary line is '5'.

Note what I have done here. Each of the 'sight lines' manufacture a Pythagorean right triangle and the inscribed circle touches the perimeter so we have arrived at the same position which so perplexed Pythagoras some 2,700 years

152

ago and which perplexed geometers an mathematicians for long centuries before being more or less resolved in the 20th Century. The confusion between the two distinct C sets cannot be eliminated but there is no reason they cannot overlap as long as we exercise due caution to avoid mixing the two sets.

This brings us to a reunion as a final factor is called up and the various figure 7's are reintegrated into a solid system. It is a simplistic maneuver which is forced by the inherent logic of the affair. Figure 8 perfects the reintegration by recognizing that the perimeter of the box is bounded on all sides by the infinity of encircling ylem. Our little shift of coordinates and change of origin does the trick

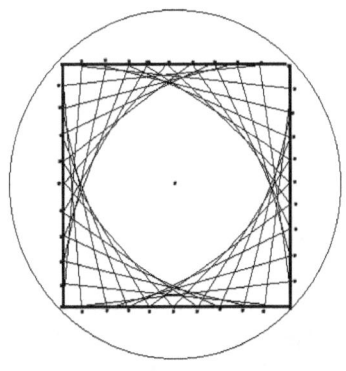

FIG. 8

We have inscribed the circle within the box. Now we inscribe the box within a circle and the paradox resolves itself automatically. The inner circle is the hole in the doughnut and the outer ring marks the boundary of the enveloping ylem. The smoke ring which is our universe is bow clearly defined.

We still have a few minor embellishments to add to our picture. Einstein, for instance, overlooked a possibility when he arrived at his 'saddle-shaped' geometry of space. He was certainly on the right track but he failed to consider the topological implications of the hyperbolic parabolid (or 'saddle-shape') and portrayed the base as being flared outward. Recurving it back is topologically identical.

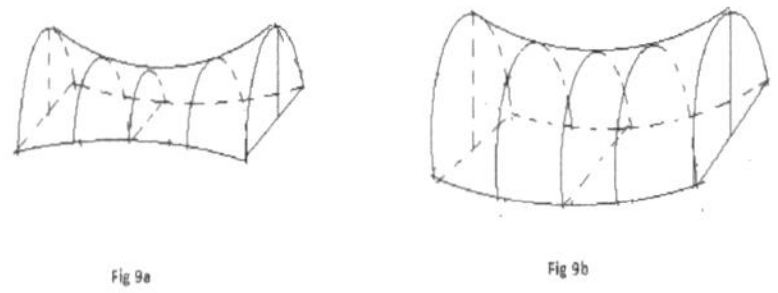

Fig 9a   Fig 9b

FIG. 9a/9b

Summarizing thus far; I have focused exclusively on the subject of light. Space *per se* is empty nothingness through which various permutations of energy are scattered. So the idea of somehow warping it is sheer nonsense, dreamt up by physicists to conceal their barrenness of imagination by manufacturing another 'Ptolemaic' universe. But a smoke ring filled with organized ylem in immensely slow whirling rotation solves a host of otherwise intractable problems without any need to postulate all sorts of arcane mechanisms.

There is no need to argue for a mysterious 'other' universe which leaks over into our universe to fill it with bosons so we can have both mass and gravity, no need to foist off a space consisting of a nothingness which consists of a mysterious somethingness capable of being deformed by the leaked mass, no need for tachyons to balance the books and no need to stray too far afield of ordinary physics in order to account for our very existence. Above all, there is no need to postulate the existence of any physical phenomenon other than ordinary energy.

Am I correct? Probably not entirely, and there is much more yet to be discussed, so it would be premature to venture more than a cautious weighing of the evidence, both for and against my conclusions.

The preponderance of evidence and logic argues that each photon of light consists of innumerable photes. These photes each possess a minuscule mass and gravity. But this is not entirely certain. It is merely suggested by the physical evidence, and no more. As the adage agrees, "to err is human, and to forgive divine" --- which excludes preachers, politicians, lawyers and many academics or critics; all of whom seem singularly averse to forgiveness'. Since I confess to being one of the fallible branches of humanity I feel quite confident I have manufactured any number of blunders; most from stupidity with a few arising from typos or simple fatigue. But equally obviously, if I knew where these mistakes are I would not have made them. My solitary alibi for making them rests in the fact that I am traveling along mostly uncharted ground where there are no guideposts to direct me along the proper paths.

This pretty much wraps up the easy part of my little essay. Now things start getting more complicated, and on this note I abandon my focus on the character of light and turn to more mundane tasks. This line has just about mined out.

VIII

A BIT OF HALFWAY MATTER

Getting to this point was fairly easy. Now things start to get difficult. Any analysis such as I have made here necessarily carries a load of baggage along with it. At best it merely points us in a given direction without adding much to our comprehension. It is a workable foundation which may be of use as a starting point but should never be taken as an ending point.

The question of how to create matter from nothing more than light remains to be discussed, and here we are really treading on unmapped ground, albeit a rather considerable mass of experimental ground has been covered *ala* the giant colliders and other techniques. These discoveries are almost invariably depicted in a 'top down' context where everything starts at the top with the completed atom or particle and degrades downward in hopes of discovering a starting point so we can start building up again.

Now stop and think about it from a broader perspective and the truth is inescapable: The best the physics chaps have been able to glean starts with a Big Bang when a cosmic egg erupted and proceeded to produce a cascade of hyperparticles which, in their turn, promptly degraded into progressively smaller and smaller particles until a few of them successfully culminated in elemental hydrogen, which then started bulking back up to manufacture stars capable of transforming their miscellaneous hydrogen atoms, which, in their turn, fused into more complex elements! It all sounds ever so pretty. It is almost mystical in its sublimity. But sustaining this picture has requires a vast proliferation of codicils which combine to produce an entire arsenal of increasingly fuzzy conclusions.

Let us begin by pointing to an egregious flaw in the presentation of physics. It is superficially innocuous but it pervades much of our mathematical analysis. Succinctly, it consists of a kabalistic faith in the divine sanctity of numbers together with an abiding shamanistic belief that if you can pin a name to an object or phenomenon you have resolved it!

To illustrate, in olden times if you fell ill the shaman would perform his rituals to decide what demon had invaded your body to sicken you. A modern doctor may decide that you have been infected by a germ. So what is the difference? Call it a 'germ' or call it a 'demon'. Both are the same. They would be the same if you diagnosed it in Chinese or Arabic. They are simply ways of saying *'Hey fella, something has invaded your body and you are sick!'* these words signify nothing until they are taken to convey some degree of comprehension concerning the root cause of the illness and how to treat it.

In short, semantics can never be an acceptable substitute for actual understanding. Now let us see where this leads us. What is the abiding sin of modern physics?

We know the Earth sports a magnetic field capable of deflecting harmful solar radiation. We also know that Jupiter has an even more powerful field, and both Saturn and Uranus seemingly also possess shields. The sun exhibits titanic magnetic flares arching a quarter of the way around its equator.

Intergalactic space is known to contain enormous magnetic clouds seemingly existing without any material support... just sitting there innocently doing whatever their thing is. The center of our own galaxy may be spewing opposing magnetic eruptions from either pole. This latter is not certain, but several similar galaxies of the same type as ours have been discovered which do exhibit the phenomenon so it would be premature to assert that our galaxy should be an exception... at least not until more is understood about the phenomenon.

This is all well and good, but now I point out that a magnetic cloud in deep space must be radiating some sort of energy or we would never be able to detect it. This directly implies one of two things, but not necessarily both; i.e., that there is some outside source such as light, gravitational energy or intergalactic debris feeding the clouds thus replenishing their energy, or that they are slowly evaporating due to the loss of energy.

A likely alternative to this may be seen in an elementary school physics demonstration of a bar magnet. Place the magnet beneath a sheet of thin paper or cardboard and sprinkle iron filings on the paper and we find one possible answer. The filings promptly align themselves along an arching path extending from one pole of the magnet to the other. And please remember that this is a two dimensional representation of a three dimensional phenomenon which effectively cocoons the magnet from outside interference. It also proves that the magnetic field has a distinct structure. It is not simply an outpouring of energy which dissipates into space and depletes a concealed cache of energy within the magnet. No doubt this is not an infinite condition, but it is at least of long duration...

possibly into the billions of years under some circumstances.

A different alternative points to those polar ejecta streaming deep into intergalactic space. There do not seem to be many of these galaxies, but there may be enough to account for the relatively few magnetic clouds in the space between galaxies, which would perhaps make it a one-time event in a galaxy's history. Either alternative is a legitimate possibility and there may be even more alternatives.

But note that there must be a persistent energy leakage from all such systems or we would be unable to detect them save as inferences gleaned from observations made when we attempt to look through these otherwise invisible fields to study the light emitted from stars or other phenomena on the far side of the magnetic field. This is what I meant in saying matters would start getting murky in Part 2.

We are dealing with many unknowns and the best any of us can do is to lay out a few conceptual alternative avenues to pursue the issue.

But one thing is certain, where there is structure there is also organization, so here we find ourselves contemplating the fact that light may be characterized in terms of its architecture. This immediately raises the question of whether all types of light also possess an architectural character, or is it merely a fluke confined to magnetism?

Of interest here, we have an apparent replication of the chaining phenomenon of light, which argues for a long succession of head to tail photes to manufacture a coherent beam of light; which suggests that each beam of light may be three dimensional just as the magnetic field surrounding a bar magnet or Earth's Van Allen belt is three dimensional.

A CBer's side band radio also argues in the same direction, though not so emphatically since radio transmissions typically exhibit a spread of frequencies.

The polarization of light virtually requires three dimensions, as do the absorption and emission of Fraunhofer lines in a spectrum. Lastly, but not least importantly, there is the odd fact that light seemingly is both a wave and a particle.

I know I have discussed this earlier but now I am approaching it from a rather different angle, so please bear with me a while longer.

Were these particle/wave aspects of light considered individually they might be dismissed as inconclusive, but when taken together the conclusion is obvious; the individual phote must possess a distinct architecture. The problem now becomes one of determining what sort of architecture is required to accommodate the features identified here?

To the best of my knowledge only one structure fits. Visualize a minuscule phote in isolation. It is perhaps 10^{-40} cm in length and has a slight curvature which is imparted by the forces implicit in the birth of universes (which will be discussed in the next chapter).

Among the characteristics of this phote is magnetism, the 'embryonic' ingredients of mass and hence gravity, with the gravity being merely a product of mass. The chaining of photes and alignment of light are products of its tiny magnetic capacity which reside within each phote and hence remain largely undetectable by our available instrumentation.

The curvature of the individual photes, when connected *en chain,* determines the polarization of the light beam. In addition, since we are dealing with the three dimensional phenomenon, it means that the beam actually proceeds along a helical track. We therefore amend our initial visualization of the photon chain to depict it as similar to an old-fashioned spring on a screen door; long, rather limp but structurally coherent as a corkscrew helix.

Now take this supple spring by one end and dangle it limply. Then exert a gently swaying motion and watch what happens. If the spring is long enough we will see a rippling effect appear.

Put it all together and what do we find?

First is a differentiation between two entirely different motions; the particulate structural motion of the chained photes, which provide the radiation pressure and enter into equations requiring mass and gravity, and secondarily a wave motion imparted by the vibration of the photon helix. In short, when we speak of light as being both a wave and a particle we are actually dealing with two separate and distinct concepts; the particle, which is light, and the vibration which provides the wave motion. It is all very simple and quite direct. The currently existing model is merely another product of unadulterated academic genius; which so often remains obstinately oblivious to the obvious in order to embrace the exotic so as to satisfy a smug hubris.   Note also that this model resolves what some exuberant physicists call the e.s.p. of photons as well as the mysterious coherency of the Fraunhofer lines in a spectrum.

There may be other problems with this model, but if so I am unaware of them and regard it as at least a step in the right direction.

There is an interesting sidebar to this analysis. Note that while the phote and its attributes may be thought of as real, the superimposed wave and its consequences are purely notional. They do not exist as things so much as the appearance of things. The waviness is real, but it is not a *drang in sich,* a 'thing in itself'. It raises the prospect of there being more such artifacts along the way. The concept is admittedly difficult to grasp, but perhaps a general grasp of my meaning may be captured by thinking of a pair of human legs. These would be indisputably real, but when we think of the legs striding along a path the strides may be real enough but they have no independent existence. They are mere artifacts attending the reality of the legs. They are purely notional. The same sort of reasoning applies to the photon change,

So do not be too hasty in confusing artifacts with reality. It may easily result in absurdities, all of which combine to leave us with the fairly simple question of how this aggregate of photes managed to transform themselves into what we fondly call 'matter.'

It is an easy, direct, operation; one made possible by the slight curvature of the phote. Consider an abbreviated chain, one consisting of perhaps 50 or 60 photes with a combined length of 10^{-39} cm. If confined within a soup of photes, all of them contributing conflicting magnetic vectors, the array may react by looping around to recurve upon itself. But the same pressures which compelled it to loop back must also torque it. As an analogy, take an ordinary rubber band, then start twisting it around until it collapses into a tightly convoluted ball. You now have the primal electron and the first step in building particles and transforming them into atoms is now complete.

IX
SUMMARY

This completes the first part of my little excursion into unorthodoxy. It is also the more difficult part, largely because so much of it ventures into enormous expanses of untrodden ground which is ignored by physics buffs.   Am I correct? Probably not entirely, but I think it likely I may have generated thoughts about elements of physics which have hitherto been relegated to obscurity because they require actual thought rather than blind knee-jerk worship of the tools available to generate raw data. But this has long been a characteristic of us human types. We take chunks of stone and chip away at them until we have manufactured gods, whom we worship. We produce dyes with which to paint our gods, which we promptly worship. We print money and worship the wealth it provides. We devise abstruse mathematics and then worship their output. We create writing and promptly start worshiping the printed page. We create beauty only to begin worshiping beauty.

And, most dangerous of all, we worship our power while lacking the faintest idea what to do with it. The things we most worship are the creations of our own hands and our own imaginings. Ultimately we worship ourselves, as the Jewish and Christian Bibles agree and the Koran supports, i.e. Satan only rebels when Yahweh/ God/Allah commanded him to worship man! And this egotistical blindness inevitably leads to egregious blunders. It also raises a disturbing question: *Since God supposedly created man in his own image, does this mean that God worships man?* I do not presume to answer this question, but it is certainly provocative.

I do not know where I may be in error. But as I used to tell my students many years ago when I was teaching, "Probably half of everything I tell you here is wrong. The trouble is, no one knows which half. Today every reputable scientist regards phlogiston as an amusing speculation. Yesterday they were praising it as the supreme achievement of modern science and a triumph of human intellect! So what will we be equally certain about in tomorrow's faith?" I feel I am probably right in what I have written, but intellectually I accept as fact that I have no doubt written some false-to-fact, perhaps even nonsensical drivel. I leave it up to others to sort out the true from the false.

Parts three and four of the cosmology will be comparatively simple, largely because it is premised on this part and deals with such minor elements the motions of particles within atoms, the origins and age of the universe, the development of planetary systems, the ultimate destination and ending of the universe, plus odds and ends of tidbits which fall between the Alpha and the Omega. I also wish to offer my apologies to the many dedicated and earnestly seeking scholars who do know how to think and *are* open to new and untested ideas. Most of them tower head and shoulders above the mediocrity inflicted upon them by a money grubbing educational system which annually disgorges vast numbers of doctors of physics whose solitary virtue is their ability to manipulate their confusers while lacking the mental wherewithal to know what they are doing but sport egos all dolled up and inflated to mask their lack of wit. On this note I write an end to Part i.

INTERLUDE

Joyce Kilmer is recognized as a great poet chiefly because of his magnificent poem "Trees". I was fortunate to obtain an outstanding blank verse ode by an unknown author but which justly merits immortality and worthy to be required reading by all who are interested in this universe.

NOTHING IS FOREVER

By

Steve Dopogney (1960-2011}

An Evergreen Hospice and Palliative Care Volunteer who was killed by a drunken driver.

Nothing is forever

Not the deserts.

They were once seas,

Not the beaches.

They always move

Not the rocks and mountains.

Erosion takes its toll.

Powerful forces

Multiplied by time

Wreak havoc on our world

In Slow motion

Nothing lasts forever

Least of all me.

Not my thoughts. Not my memories.

Not my words.

They will be nothing.

But there is forever

And there is continuity

Whether men remember or not

When they pause and reflect

On what has passed before.

The fact of the matter

is matter is me.

*I cannot be created or destroyed. But only changed
from one form into another*

So the very atoms of my existence

Will continue in this world

Until this world ends

hereupon they will be released into the universe

As matter or energy. Or particles or waves.

But that bit of energy, that bit of mass

Was once me

And therefore I will stay

In this world, of this world

Unseen, unrecognized

But still here, disparate parts

They spring anew and oblivious back into circulation

Of the water of the hydrosphere

Of the carbon in the food chain

Of the energy used to produce

A sound

A city

A new person

And some small bit of me will still be useful

Will be part of the world

Part of the very matter of their existence

Even though they do not see me

Do not feel me

Do not remember me

I will always be there.

PART TWO

XI

BANG!

As I mentioned toward the end of Part 1, it requires a special genius to remain obstinately oblivious to the obvious, and many physics chaps have proven themselves particularly adept at demonstrating this truth. For decades they pondered, and even exulted, over the paradox of light's seeming ability to be both a wave and a particle without even once considering the possibility that they were dealing with two distinct elements, one being a physical corkscrew architecture, the other being a notional wave deriving from the vibration of the flexible corkscrew. Now where does that lead us?

Imagine yourself resting atop a bar magnet, now consider that this is merely a two dimensional simulation of a three dimensional coil spring. (A slinky with one end stapled to the ceiling works the same, so think of it in that fashion if you will.} Either way this may be taken as representing the physical being of the photon. The depiction includes the mass and gravity inherent in every phote, and when detected by humanly devised instrumentation are regarded as 'Compton' waves since they appear on a flat screen and thus seem superficially identical to a wave profile. Now we crank in the vibrational mode of the emitter as it reacts to the thermal excitement to create an oscillation in the spring and voila! We now have an immaterial, purely notional, wave motion superimposed atop the material spring. Case solved…and without recourse to paradoxical or arcane subterfuges meant to conceal ignorance. It is simplicity itself; which makes it off limits to modern theoreticians who revel in creating complex mazes and refuse to make simple statements so long as arcane substitutes are available to awe the plebs.

The whole affair is amusingly reminiscent of the medieval church philosopher Peter Lombard, who was known among his peers as "The Labyrinth" for his ability to reason around in ever widening circles (much like those among mathematicians with their "ring" cycles) spinning from page to page before getting completely lost; at which point, voila! he would wrap it up with a conclusion which defied comprehension, thus convensing his rivals that he was a profound master of his trade!

*NOTE: I hope the reader can decipher the above sentence. I sought to make it coherently incoherent but I am uncertain if I succeeded.

But this is only part of the story. It makes an excellent introduction while leaving many questions unanswered. What of the dark lines in a spectrum? What about the lateral connections which make light appear coherent at a distance? What of the titanic magnetic clouds deep in interstellar and even intergalactic space? What of the dark matter which permeates the universe? How does light get transformed into electrons and particles? And most of all, how did the universe get started?

These ought to be enough 'what's' and 'how's' to last us a while, but if not, maybe we can dream up a few more.

Trying to consider each of these questions individually would be both confusing and involve unnecessary redundancies. Accordingly, suppose we choose to start at the beginning of the next stage of our cosmology, i.e., manufacturing a universe. Many, if not all, of the what's and how's selected above will explain themselves in the process. May we dare to call it a new beginning? That sounds so pretty albeit not quite right. So why not think of it as the second creation... or perhaps merely as recreation? Either of these sounds acceptable.

Start by visualizing a vast, endless void; an empty cathedral lacking walls, floors or ceilings if you wish. Then stop to examine the meaning of 'endless' and put it into Cantorian terms. To an external viewer this void may seem a simple pinprick on a sheet of paper. To an internal viewer it may represent tens of billions of light years. It all amounts to the same thing. From one direction you are a scientist peering down at the universe through your microscope. From the other you are a tiny bacterium scientist stuck in a petrie dish looking up at the enormously magnified eye at the microscope. Either works.

Being empty there is an eternity of time, which is felt as no time at all because nothing ever happens which can connect two events. Lacking a concatenation of events an internal viewer could have no sensation of the passage of time even though an external viewer might see an entire universe flash into view and live out its existence in a tiny fraction of a second. It is genuinely relative to the position of the observer and there are no reference points available as anchors.

This is our starting tableau. For our purposes it is probably best to adapt our thinking to a human scale of hugeness, but we ought never to fall into the intellectual trap of believing it must be titanic to outside observers. This space is not exactly empty. Far from it in fact. Adrift in this vast nothingness is a sea of tiny, disconnected and unwaved Photes. The mean density of this sea is purely speculative so no point would be served in seeking to estimate it.

Additionally, the origin of this ocean of photes remains a mystery beyond the scope of any human mind, i.e., what is on the other side of nothing and where did it come from? When did it all start if there was no time, etc. If you wish to throw in a god, then be my guest, but remember to ask where did this god come from? In brief, these are idle imaginings best left to theologians, charlatans, and mountebanks, all of whom pretend to a knowledge they, in fact, lack. It would also be a mistake to think of this ocean of photes as stagnant, Far from it. The mass of the individual phote is minuscule, ergo the slightest touch from any other object also possessing mass must send it caroming off at whatever is the rest speed of objects in the ylem. Since the ylem is posited as containing only photes (at least in the original configuration) this must imply a formidable recoil speed, and since it would be one phote colliding with another they both will recoil accordingly at speeds approximating c in this environment (which need not conform to c in our universe).

Such collisions will be rare no matter what the start-up density of the ylem may be; they simply are neither large enough nor massive enough to interfere with one another on a regular basis, but occasional interference will occur. When they do occur one of two outcomes is inevitable; i.e. the mutual recoil just mentioned, or chaining. The latter occurs when one phote impacts another at precisely the correct angle to allow the conjunction of their magnetic axes. A rough analogy to this process occurs when a bacterium which normally is solitary occasionally encounters a similar bacterium and unites briefly in order to exchange DNA. The analogy is far from precise and is seldom observed, but mitosis has frequently been remarked on and just about every bacteriologist is acquainted with it. Either way, the ylem will ultimately come to possess a quantity of linked photes. It need not be a huge number, but in Cantorian terms it may be regarded as infinite without fretting over the matter. What is more significant is the fact that any phote chains manufactured in the ylem will be unwaved and thus undetectable by existing instruments. In other words, their motion will be lineal and conform to Cantor's c set of straight lines.

Time remains unknown within the photes of this ylem since the ocean of turbulence has about the same lack of eventfulness as the droplets of water in our ocean. They are random and non-sequential.

Being this turbulent it is predictable that the density ratios within any given region of the ylem with respect to adjacent regions will be continually varying, with some regions becoming markedly denser than others. With this increased density comes an increase in the frequency of phote pairings. Of greater significance here is a realization that the denser the ylem the greater the incidence of pairings as well as a pronounced increase in the localized self-gravitation. Allow it to proceed far enough and it starts drafting the untapped ylem throughout the region. Ultimately it reaches a critical mass which then explodes to create a universe.

The scenario just described is not original with me. It was first advanced by Gamow in the

1930's (who also contributed the Nordic word ylem [Chaos] to the scientific lexicon). It was shortly thereafter discarded and cast into the scientific dust-bin equivalent of ylem obscurity reserved for failed theories. Interestingly, the basis for its rejection lay in the fact that according to Gamow's thesis any explosion must develop at the earliest possible moment, just at the instant criticality is reached. This gives rise to the bare minimum blast energy, only a little big bang, which would provide insufficient energy to create the super particles needed to cascade into the more familiar protons, nucleons, electrons, neutrinos and the others inhabiting what has been referred to as the "particle jungle."

This left only the 'Big Bang' hypothesis still standing; or so it would seem. According to this model our universe began as a titanic eruption in localized space...a 'cosmic egg' or volcano which tore through the 'fabric of space' with unimaginable violence and began expanding, originally at supraluminal speeds in all directions but later slowing to c.

This is an interesting idea in its own right because when translated it turns into the astonishing declaration that the expansion speed is greater than infinite!

Be that as it may, in its original apparition this super bang was assumed to be several thousand parsecs in diameter but over the decades it was calculated that not enough energy could be generated from a cosmic egg of so large a size, so it was progressively shrunk until in today's physics it is mere centimeters in diameter.

Current theory does not really address the question of what force would be capable of overcoming the enormous gravitation of our universe if it were compacted into a robin's egg sized (or smaller) volume of space, so I leave the matter there. Suffice it to say that a diameter of less than 3 cm is among the more recent candidates for the honor.

Big Bang or little bang; it is easy to overlook the psychological implications separating the two approaches. The Big Bang is representative of a theological "top down" approach where a stupendous eruption manufactures giant particles which then break up into a cascade of smaller and smaller particles until they finally arrive at protons, electrons, nucleons, neutrinos, etc. At this point they start clumping together to form stars, which in turn take these minimal particles and start fusing them into the heavier elements we see about us today.

A little bang reverses this. With this the initial creation erupting from the ylem would consist of protons, electrons, neutrinos, etc and go from there, thus avoiding the downward cascade of hyper-particles. Although it was not present in the initial presentation of Gamow's theory, the hyper-particles generated by giant colliders are simply ephemera, rather like the debris left on a floor after a big glass ball is sent crashing into a window pane... i,e, they are simply slivers of glass having a momentary coherence but still in the process of disintegration.

A not so minor added problem here is the apparent near absence of antimatter in our universe. According to theory a Big Bang explosion should be indifferent to the handedness of the particles being manufactured, yet ours is definitely a 'left handed' universe. Theorists have been able to arrive at only one potential explanation for this anomaly, and that by pointing out (correctly) that nature is ordinarily a pretty sloppy affair. Suggesting that a perfect one-to-one correspondence between left and right handed particles would be contrary to experience and highly unlikely, they have theorized that the initial explosion perhaps manufactured 51 percent left handed particles versus 49 percent right handed ones, with the result that 98 percent of the whole would be destroyed and only 2 percent would remain to occupy the remaining universe.

Now comes the kicker, which seems to have been pretty much overlooked in the calculations. If our universe is no more than 2 percent of the total output of the bang, then the initial bang must have been fully 50 times more powerful than we are presently calculating, which in turn further reduces the diameter of the cosmic egg. By now it can only have been about the size of the period at the end of this sentence.

Parenthetically, it also raises the question of what happened to the energy spent in the matter/antimatter destruction! After all, we are postulating energy to demolish energy, therefore the output must be light, which is itself energy and therefore merely a reversion to the ylem --- an interesting idea, is it not? It might easily be thought of as a submicroscopic yo-yo effect.

I may be a few decimal points off base here since I have merely ball-parked my figures, but it hardly seemed necessary to devote much time to the matter. The Big Bang is a seriously doubtful model and its problems are not ended. Even allowing for all this, there are other logical paradoxes to be found in the contemporary theory. Chief among them is the fact that a spherical explosion compels a lineal geometry which works well enough along any radii extending from the core to the perimeter of the expanding universe but cannot overcome the distortions introduced by adjacent lateral galaxies.

Logically the big bang idea must posit an abrupt eruption with an approximately zero time before completion. The relationship between gasoline and nitroglycerine explosions is a useful analog. A big bang implies a nitroglycerine style event where the total energy is generated in a few tenth of a second. Given this it is evident that the resulting universe will be spherical with all the ejecta concentrated around the surface of the expanding sphere. The often employed balloon depiction is approximately correct here though there may be an occasional laggard galaxy which has not kept up with the rest plus a few which may have led the pack.

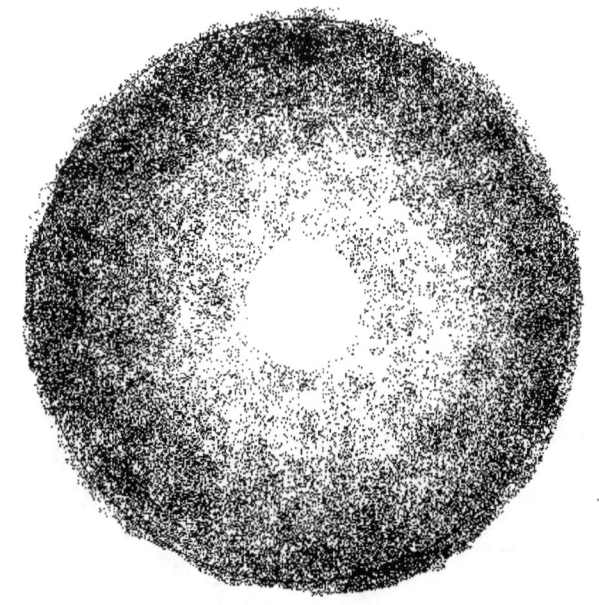

FIG. 10

Now consider what we see when we turn our telescopes heavenward. Nowhere do we see any evidence of the layering we would expect were this true. This would seem to provide even more evidence that space is curved and we are in fact looking around the balloon's surface!

Now we may go back and try out another hypothesis; one where the time element is more diffuse and the exploding universe takes a little longer to develop. Admittedly this contradicts the requirements implicit in the big bang model where a strongly focused detonation is postulated, but a bit of added reinforcement is appropriate.

In this universe we see newly ejected of splatter galaxies strewn all over the place in disorderly array.

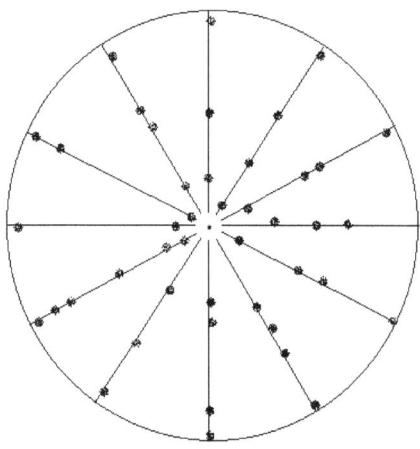

FIG. 11

It works for the expanding balloon analogy only if we presume that everything is departing the origin at a common speed so it is all surface with zero depth. This would seem inherently unlikely. Doppler measurements would be unambiguous were either of these the case and this we do not see. So we are left with an unsatisfactory theory which we seem devotedly, even resolutely determined to twist around to make work somehow.

In brief, there are simply too many conceptual and logical flaws implicit in the rationale underlying the whole Big Bang approach.

So we try another tack. Admittedly, we are being redundant and mostly flogging a dead horse in spending so much time on an already discussed matter, but whenever a radical evaluation of established academic dogma is called for it is wise to point out how the proposed changes come from every which direction. The path I am presenting resolves so many problems and eliminates so many wild hypotheses it ought to convince all save the most obdurate mossbacked Ptolemaicists of its correctness

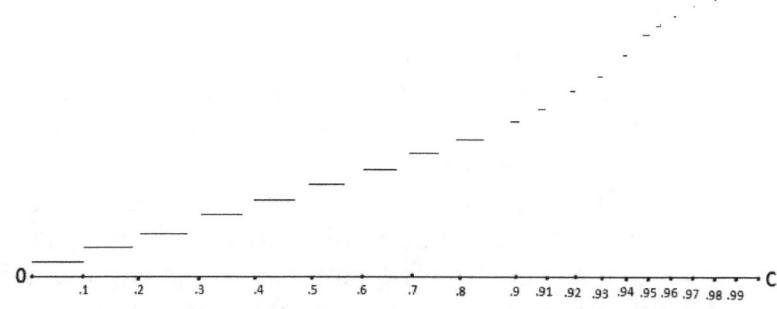

FIG. 12

Figure 12 should be visualized as a single line representing speeds ranging from 0 speed at the left to >c at the top. The intermediate divisions renote speeds upward at tenths of >c since foreshortening accelerates markedly after an object passes, 9 >c the staggered lines above banked about each line indicate the progressive foreshortening of a spaceship over as it accelerates.

Preliminaries out of the way we can turn our attentions to the logical implications of this foreshortening. And these are somewhat at odds with current scientific mythology which assumes the gravitational forces would literally pull you apart when you get too close to >c when you are caught in the coils of a black hole.

We have no idea how fast the Earth is moving in space. Nor have we any idea in shich direction. All we can know are speeds and directions relative to other stars, galaxies and planets but existing theory would have us believe that the mass of a firmly anchored steel block will exhibit minute changes as Earth orbits the sun which orbits the galaxy which orbits the universe. These changes have never been detected and we are not conscious of them if they exist. By itself this is not a fatal objection but it may be suggestive in establishing minimum and maximum parameters and/or mechanics, e.g. is our space craft being pushed or pulled toward >c, or perhaps a combination of both?.

The logic here may be annoying but it is still logical. It tells us that Earth and our solar system may actually be crawling along at something like 0.6 >c and foreshortened accordingly without our ever noticing it! This would argue that no matter what the speed we might be moving we will not notice any difference in our size and shape and the science fiction idea of being torn to pieces and reduced to our component atoms simply does not work. There are compensating factors at play... or perhaps contemporary science has simply gone off the track?

So what are the rules in this game?

Play the science fiction game and visualize a spaceship 100 meters long and 30 meters in diameter which is approaching us at about >c speed. It matters not what our speed relative to the universe might be. So far as our perceptions go all we can imagine is that a fying saucer is looming on the horizon. Abruptly it flips on its side to avoid a collision and we get a brief glance of its profile. We cannot see it because our senses do not react that swiftly but our instruments can and a later review allows us a juster interpretation of what we have seen.

Given this we may predict that the science boys will declare it is something akin to the Klingon cloaking device while the government politicians and bureaucrate will blame it on our faulty interpretation of a mere weather balloon and the theologians will declare it a harbinger of the last days sent to them by their god as a warning to increase their tithes... and only a God knows what the news media chaps will come up with.

To complete my little digression, during the flicker of time while we see the alien craft edge-on it vanishes! It vanishes because it is so foreshortened it is paper thin. But to the aliens inside everything is quite normal They are long, short, fat or skinny the same as usual and we are the ones who are distorted by the speed of our movement through space, and since we are moving at only 0.6 >c their instruments and our foreshortening would seem right at 0.4 >c and ironically, if we were perfectly still in the universe they would miss us altogether because we would be paper thin in every direction!

We are back to Zeno's old paradoxes. 3,000 years and we haven't learned our lesson. You simply cannot measure infinity. You can utilize and measure segments of any infinity but not the infinity itself. So here again we have proof that >c cannot equal C.

By implication no matter how we try to arrive at this black hole our instruments and perceptions stubbornly refuse to accept it. Light speed is unchanged , the hole does not get any closer and we are not conscious of anything different! That is the ultimate secret of relativity. Einstein had the right idea all along. He simply did not pursue it deeply enough.

Now note that there is another element to this. We take a particle and accelerate it in a huge cyclotron and recall my irreverent scoffing at some of the antics of our new breed of delvers into the unknown. Start by looking back a few centuries and see what Galileo had to say, which chances to be highly relevant in this day and age.

$K=1/2MV^2$. The kinetic energy of an object is equal to its mass times the square of its velocity. This is a beautiful equation, elegant in its simplicity, profound in its implications and often ignored or lost in the welter of arcane circumlocutions of modern descendents of Peter Lombard who revel in the pursuit of complexity.

Current theory argues that light speed is infinite, i.e. >c=C. But the same theory rejects the idea that mass can make the grade since that would imply situations where $K=1/2 MV^2$ should be written $K=V^2$. In short, no matter how we choose to look at it or ignore it, >c cannot equal C.

By extension, when we accelerate a particle to 'smash' an atom we are literally pumping energy, i.e. mass into the particle. By the time we are finished and the particle collides with the target it may possess the mass equivalent of a freight train smashing a glass jar and what we get on our camera plates are the ground up slivers of glass.

Translate this to a nascent universe. Where this energy came from or how it was generated, or how it was confined prior to going bang is necessarily speculative. This is the first flaw. The second flaw is probably more significant since it relates to much astrophysical speculation. This is the naïve belief that everything has to be taken to the utmost limit. I illustrate through a theological analogy since it is evident that much of astrophysics, including my own, is tinged with theological perspectives (which is why I have made all of my perspectives explicit). Start with the supposition that a god actually does exist and that this god manufactured the universe at a given date which we have calculated to be a Sunday with today being Thursday on a theological calendar which ends with the destruction of Earth at midnight on Saturday. This is not too distorted from much primitive Judeo and Christian speculation that dates creation of the world at 4004 b.c.e. and pursues the birth with endless predictions of the day of our demise.

Please note that nothing in this week long existence precludes the actual creation from having happened on Wednesday, together with apparent potential reasoning calling for a creation extending back to Sunday; which inevitably brings us to the logically flawed Cartesian dictum that god would not lie to us. But there is no dictum which prevents us from lying to ourselves. Blaming some deity for all that ails us is a tremendous burden to place at the feet of a god. It amounts to ordering god to behave according to our faith; which does not seem altogether fair on our part; and perhaps even hazardous to our health in an afterworld. It may well be that a god possesses a quaint sense of humor and merely enjoys observing us in our theological antics; a sort of theological stirring of the ant hill if you will!

Summarizing, science may calculate the energy requirements needful to manufacture our universe in a Big Bang, but there may be other ways of arriving at the same end which do not require so much energy being generated in a single instant from a minuscule point. There may well be shortcuts we have not explored and it ill behooves us to take it for granted that we have exhausted all the alternatives.

In brief, the Big Bang model is interesting, and may actually be correct, but we ought to bear in mind that, being human, we may be making a serious mistake when we ignore all other possibilities.

So suppose we take another look at Gamow's model and see where it may lead us. Begin by seeking flaws in his initial presentation, starting with the most obvious one. Whether consciously or unconsciously Gamow started with the premise that first there was this ylem, then it contracted until it reached criticality and thereupon promptly exploded. So now there is a universe and no more ylem.

Note further that the model I have used here supposes that the universe follows from Cantor's analysis of an infinite size, therefore our universe may well be a localized event. Accordingly, Gamow's secondary premise does not necessarily follow from the primary premise. It is deficient in logic and adds a needless layer of complexity to the model. There is, in short, no logical reason to suppose that the whole of the ylem was involved in the explosion, and recent research (about which more later) actually suggests that it represents only a minor fraction of the whole. This conclusion holds even if Cantorian transfinite mathematics are ignored. In brief, Gamow's secondary hypothesis arguing for the extinction of the ylem must be discarded.

We may rephrase Gamow to specify that turbulence theory in an otherwise empty medium is inevitable (probably following the Kolmolgorov turbulence pattern analyses). Instead, we have a localized event with the remainder of the ylem unaffected and today is manifested as dark matter.

Now consider the fact that a liter of gasoline contains more energy than a liter of nitroglycerine. The reason nitro is so dreaded rests in its ease of detonation and the power of the explosion. Gasoline burns more slowly and releases its energy over a period of time, so it is thought of as somehow being 'tame'. The technical jargon to define the difference between nitroglycerine and gasoline is 'brisance'. Nitro is a high brisance explosive while gasoline is low 'brisance'. When it comes to manufacturing universes the Big Bang hypothesis is ultra-high brisance where the Gamow model is low to moderate brisance. Were this the entire story we would be compelled to conclude that no little bang was possible and the Big Bang would be the sole remaining alternative. But suppose we fiddle around a bit and study the implications inherent in an ordinary bang.

I have begun by introducing a turbulent, infinite sea of mostly unwaved photes, which collectively make up the ylem. The turbulence implies localized variations in the density of these photes with occasional regions approaching criticality. Given this, it is to be expected that the regions surrounding this looming criticality should also be regions of higher than usual density, otherwise the high density regions would rapidly bleed over into the lower density ones. With some low density regions being gravitationally attracted by the high density regions there will be a tipping point where outgassing will largely cease and we will have an approximately globular volume of space where the phote density ranges from high at the core to negligible at a distance of possibly as many as several dozen light years (if looked at according to our perspective. What it might be to our hypothetical researcher studying the eruption through a microscope is another matter).

I do not attempt to evaluate the nature of the explosion but neither do the proponents of the Big Bang evaluate the nature of their blowout, however it is evident that the energy of a low brisance explosion is many magnitudes less than required for the Big Bang to do its thing. Conceivably a fusion chain igniting H-bomb would suffice for a low brisance one, but it would not in itself correct for the problem of manufacturing the necessary particles. Still, a closer look at the prospects of a low brisance event might help resolve a thorny physics issue.

When I was a lad I recall seeing newsreels where our battleships fired their mammoth guns. If watched closely you could see the smoke ring boiling from the muzzles after each shot. In a way this same effect would occur when the ylem's dense core commenced its detonation, though the analogy is far from exact. Looking in just one direction, call it north for clarity's sake, we would see a titanic smoke ring of enormously stressed photes erupt from the core of the ylem sphere and then proceed outwardly while expanding at perhaps a 45^0 angle.

This smoke ring would be parent to our universe.

But this is not the end of the matter. We are considering a continuing series of events happening in an otherwise empty space. The ylem still exists, even in the regions being traversed by the explosion. Precisely analogous is the behavior of a smoke ring in a terrestrial laboratory. When the ring is generated it emerges into our atmosphere and travels through it. But it does not co-opt this outside atmosphere into itself. It makes its way through with minimal effect on the normal environment, hardly disturbing the ambient air. The same applies here. The smoke ring which is our universe moves through the ylem 'atmosphere' in identical fashion. Call this ylem 'dark matter' and the parallel becomes even more exact.

Capping the situation, the ylem moves in straight lines but in every which direction, just as gas molecules in our atmosphere would move in straight lines were there no centralizing gravity source to impose a curved trajectory. Recent studies have reputedly proved that the dark matter substratum in our galaxy obeys a flat geometry rather than a curved one... which provides a serendipitous support for this model because every smoke ring is a torus, which just happens to conform to Einstein's curved saddle-shaped geometry in every respect save that rather than recurring outward as in a saddle it curves inward as in a doughnut. Topologically there is no difference between the two. In other words, the torus imposes its geometry upon its components while largely ignoring the medium through which it is moving.

This may best be depicted by a comparison of the two alternative models of the space curvature. Figure 12a illustrates the Einstein model while Figure 12b depicts the model arrived at from my interpretation of the evidence.

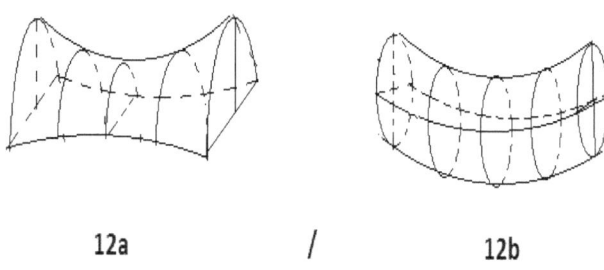

12a   /   12b

FIG.12a/12b

Now think back to the final diagram of Part 1, where a modified version of the tachyon hypothesis wound up circling endlessly within a boundless infinity. Surprise, surprise! This is precisely what we should anticipate in a toroidal universe!

The little nudge which sent me off in this direction has more ramifications than I first imagined and is now tending to confirm a smoke ring universe within which we exist.

In a manner of speaking the condensing ylem might be considered a cosmic egg and it may be more convenient to think of it in that way. It floats suspended in a rarefied soup of photes which collectively comprise the ylem (precisely as droplets of water comprise an ocean). When criticality is reached and the core of the sphere ignites, a new set of consequences ensues. Most importantly, Newton's law of equal and opposite reaction comes into play. The same eruption which manufactured our left-hand smoke ring explosion must also create a right hand smoke ring of roughly the same intensity, so a *second universe* is also created, only that one is heading off in the opposite direction. In other words the creation of one universe automatically manufactures a reciprocal one. Our universe i.e., the left handed one, we call the 'normal' one. The reciprocal one would be a right-handed, antimatter universe, so there is no longer any mystery in the scarcity of antimatter in our universe.

Nor does the matter stop there. It is not a case of the ylem splitting in half and manufacturing two opposing universes. Like a die in a crap game, smoke rings may erupt on every side, with six rings emerging from the debris… six distinct universes at right angles to one another, all launched from a single event. Of course not all of this is inevitable. Nature is a pretty sloppy affair and the distribution of the exploding ylem may not result in anything more than a 'blow-out' in the ylem, where the exploding material simply dissipates back into the empty. In other instances perhaps only two reciprocal universes are manufactured; or possibly just four, but six potential universes must exist in any event.

Next comes a little trick of nature. It requires three dimensions to fix an object in space... plus a fourth to fix the object in time. A smoke ring boiling up from the ylem core sphere automatically imposes the rules of ring formation, i.e. a developing ring rotates from the inside outward. It is merely an ordinary cavitation effect, only king-sized. By extension, the rotational orientation of our universe requires a left handedness as standard while the reciprocal universe boasts a right handed orientation. But the time dimension remains undisturbed in both since it is merely a useful technique for arbitrarily describing the interval between events (I know this is oversimplification, but essentially 'time' is a purely notional phenomenon which merely indicates the 'whenness' and develops starting from the origin of the little bang and embraces all emerging universes as one.

As for the four radial universes which may or may not be created by the same event, a bit of added explanation is in order. It requires 3 lineal dimensions to fix an object in space, ergo, our universe may be termed 'alpha' three-dimensional. But the reciprocal universe requires three opposed dimensions, so it would best be described as 'beta' three-dimensional. This is all pretty straightforward. Things become more complicated when we start to speak of the lateral four universes, which may be labeled 'gamma', 'delta', 'zeta' and 'eta', with each of them boasting three dimensions of which two dimensions are duplicated with one or another dimension alpha and beta (one in each but in different order). Here we discover that a solitary extra dimension suffices to fix them in place... a single added dimension for each means 4 more dimensions to account for here. Now let us do a little counting. 3 dimensions for our universe, 3 more for the reciprocal one, and 1 each for the potential four lateral ones. Total 10 notional dimensions, with time comprising a common 11th dimension. This corresponds to the prevailing multidimensional geometry currently in vogue. Admittedly there are a few ambitious types busily promoting a dozen, two-dozen or x-numbers of dimensions, but these are pretty questionable and may well represent no more than convenient parking places for mental exercises. But please note that all of these are purely notional. None may be characterized as real objects. They are merely descriptive words intended to indicate their separateness, hence any efforts to objectify them will be exercises in futility.

A little caveat is in order here. We have progressed from a Big Bang to a little bang, but there is a further possibility to consider. Admittedly it is a remote scenario but it is conceivable we may be able to create a universe without employing a bang at all! Start by visualizing a turbulent ylem with small phote chains meandering along in the mixture (possibly relict of some earlier universe). They sideswipe one another from opposing vectors to create the equivalent of a hurricane or tornado, which then gathers strength as it plows through the disorganized ylem. In this model subatomic particles are created as a spallation effect while the rotation of the tornado imparts the curvature to the resulting spin.

Do I believe this? No. I considered it and wound up feeling that entirely too many assumptions and unexpected discoveries would have to be made for it to be viable, but I concede a remote possibility here. So far as multiple dimensions are concerned however, it really makes no difference. Eleven is still the total. Two more little caveats are in order. First is a reminder that it is not necessary for any of the four lateral universes to form for these added four dimensions to come into existence. They exist in the abstract whether we like it or not. In the everyday world of engineering these dimensions correspond to the concept of "degrees of freedom" in mechanics so there is nothing new in postulating them. My only credit consists in having arrived at this model independently. I owe it to Walter Currah, a mechanical engineer, who pointed out that I had unwittingly stumbled onto a centuries old everyday engineering concept which seems to be beneath the dignity of the astrophysics theorists to notice or assign credit to a bunch of ordinary, plebian artisans.

The second of these caveats is perhaps more interesting. In our laboratories we routinely manufacture antimatter. In nucleonics we find six different types of neutrinos, six of bosons, etc. In current theory the reasons for this seeming habituation on the number '6' is obscure. My model for the manufacture of universes removes the confusion and clarifies the process. When we generate these particles we are unknowingly replicating the conditions which gave rise to the universe, only on a far smaller scale absent the torqueing impetus of the outside universe, thus without replicating the six degrees of freedom and therefore absent the ultra-short lived ephemera. Pursuing the analogy to an ordinary smoke ring, when one is formed in a laboratory (or by an adept smoker) and set out on its way, its axis of motion in space is unchangeable. In this respect it is rather like a gyroscope. Its rate of motion along that vector is also essentially unchangeable. In other words, it has a strong resemblance to rotation of earth's own air and water currents, which tend to reverse their spin direction above and below the equator. If a laboratory torus sets off along an arbitrary 200^0 vector at a rate of 10 cm/sec, the density of the atmosphere through which it is moving makes relatively little difference. Rather than changing its vector or speed it increases its rate of expansion and dissipation as it travels along. But though its ingredients do not alter, its physical size grows in precise relation to the density of the medium through which it is moving, e.g. if the medium is utterly void it will pretty much persist forever.

If the density is assumed to be around 0.01 it will enjoy a slow but steady expansion. Double the ambient density and the expansion rate will increase accordingly, etc. I lack the data to suggest what the exact numbers might be, and to the best of my knowledge no one has ever tackled the problem so I leave it inexact and go to other matters, except for a pair of notes. Number 1: While the tracking speed of the torus through the ylem is pretty much an invariant, the rate of expansion of the torus is variable. In short, it gets steadily fatter as it encounters the ylem photes. And this is highly variable; an automatic consequence of unbridled turbulence. In regions of negligible ylem density it will grow only slowly, but when it passes into a region of high density its growth rate increases. At some point the torus becomes so attenuated its coherency is lost and the smoke ring dissipates entirely... which affords a different perspective on the ultimate future of our universe than the most commonly held one.

A consequence of this is the invalidation of the assumed expansion rate of the universe as a measuring stick for dating evolution from the time of a supposed Big Bang. There is no mystical 'cosmic repulsion force' driving this increasing expansion rate of the universe. It is merely a measure of the growth vectors of a smoke ring. It is likely that a period of slow growth will be followed by an accelerated growth when it enters a region of higher density, followed in turn by a pause in acceleration. As with caveat number 1, where it might remain stable until it again arrives at a region where the density picks up. This argues decisively against the accuracy of the Hubble Constant in measuring the age of the universe. The present estimate is entirely too small and the actual age may well exceed 35 billion years, roughly three times that of the present estimate. I shall return to this scenario in the next chapter, where it will be shown utile in considering later development of the universe.

An added complicating factor is the likelihood that the ylem substratum of our smoke ring will include random density variations in much the same manner as our atmosphere exhibits random variations. If correct this would argue that smaller regions of our smoke ring are growing at slightly differing rates, thus distorting the apparent Hubble age even further. The next issue is the manufacture of matter, which takes us closer to home and gives us a new set of problems which must be considered. Truly, the solution to one problem usually leads to several new problems. This is frustrating, but it is also quite interesting. With everything solved and done with this would be a distressingly boring universe. Perhaps we can add to the confusion by considering a few more unremarked and presumably unrecognized alternatives.

XII

A SMATTER OF MATTER

So now we have a pre-universe as well as a universal birth tucked under our belts. But the original objection to the energy problem remains. A little bang simply does not generate enough energy to manufacture the hyper particles necessary to decay into the ones we get in our giant colliders, so where does the matter comprising our universe come from?

In this I am not speaking of the heavier elements. The focus at the moment centers on those exotic little critters we choose to call n*ucleons, electrons, neutrinos, etc., leading up to the simpler elements such as hydrogen, helium and lithium.* Lacking these we also lack matter. So how it occurs must be dealt with. Once more we take a brief look back to the 1930's, an era when brilliant, and intellectually non-constipated thinkers felt free to explore all sorts of ideas without trepidation or fear of ridicule from their compeers.

Despite all their brilliance and freedom of speculation they allowed a tiny little philosophical error to perfuse their computations. And this tiny error has both misdirected and contaminated astrophysics ever since. The error goes back to the 1800's and in large part may be laid at Georg Cantor's feet!

This must come as a surprise in view of my respect for his work in transfinite mathematics, but in extenuation he was focused more on the manipulation of C-sets than their applications in universal physics. Now we must look to the error.

Consider an object having zero mass. It is adrift in empty space at a speed of zero. Now a butterfly flits by and the tip of its wing touches our massless object.

What happens?

Given absolute zero mass it can offer only zero resistance so the target object recoils at infinite speed. No other solution is possible. Since light was believed to be massless the inevitable conclusion was that the speed of light must equal C, therefore >C = C. Q.E.D.

Given this as a firm mindset, a few 'tests' of the concept that light speed equaled C were performed, all of them resulting in speeds hovering around 297,000 km/sec. Later tests centered on the likelihood of there being 'slow' light, or even frozen light. These gave null results, ergo light is truly infinite.

Somewhere in all this a logical alternative was overlooked. Cantor's rationale was correct, but it only applies to light if light is utterly massless.

But suppose light possesses a minuscule smidgeon of mass?

In this event it will recoil at a speed slightly less than C! Now we discover an entirely different scenario lain out before us. If light is massless then it can have no kinetic energy, ergo when light strikes a mass it can only be absorbed or bounce back, so where does the observed radiation pressure of light come from? How can we have a Poynting-Robertson effect? Why is their radiant heat? How can I get a sun tan or my wife a sun burn?

Everything I mentioned here requires a transfer of energy, but if light is utterly massless there can be no energy to transfer!

This left the physics chaps in something of a quandary until they came up with the particle/wave model, and *voila*, problem solved! When light was just chugging along it was a wave, but when it hit something it was a particle.

God was in his heaven and all was right in the world, thanks to our magnificent scholars.

With the model advanced here, which calls for a real particle and a notional wave the problem vanishes, taking with it whole avalanches of determedly Jesuitical reasoning in its passing.

Ever hear of Hannes Alfven? Few people remember him today, but he was a Scandinavian physicist who, in the 1930's, briefly achieved international prominence by virtue of a theory which explained the differences between the planets of the solar system in terms of their magnetic susceptibilities. Without going into details, he proposed that solar magnetism attracted the heavy metals of Mercury most strongly, thus causing the planet to form most closely to the sun. Then came Venus, and Earth, etc, until the solar system was complete.

The whole idea was quickly shot down because actual observations proved it could not happen in the way Alfven proposed so it receded into the never-never land of physics alongside the titanic coal pile of Bickerton's sun (Bickleton was a New Zealander astronomer of the late 1800's who sought to explain the solar radiance by assuming the sun consisted of anthracite coal and the planets were created when the sun experienced a grazing collision with another star, with the planetary material being the cast off debris of the collision). Even at the time that idea was solidly discounted by astronomers everywhere.

Briefly returning to Alfven, he should be accorded a place in astronomical history books as the father of progress in the science of electromagnetics. Over the past half century. Magnetohydrodynamics has become an established discipline. We have experimented with mag/lev railroads, the U.S. Navy is testing a rail-gun which magnetically accelerates ballistic projectiles to speeds of several thousand km/hr and with ranges of 160 km or more. Electromagnetic forming has successfully compression-shrunk ordinary U.S. quarters to a mere fraction of their usual sizes and laboratory amperage has reached 600,000 amps. I have not verified this but there have been reports of electrons being created under 50,000 volt discharges.

I remain skeptical about the latter until I learn more about the protocol here and suspect that any electrons were not created but instead were scavenged out of particle debris. Such things have happened before and will undoubtedly happen again so a modicum of doubt must remain.

Capping this whole affair has been the recent, confirmed discovery of enormous magnetic clouds in remote regions of intergalactic space.

This last discovery ought to have set off bells and whistles among astrophysicists everywhere but strangely there have been remarkably few who have risen to the bait in the form of direct speculation with respect to its potential role in the origins of the universe. I suspect this stems from the enormous compartmentalization which has come to afflict so much of our brave new world. "I am a specialist in faint blue stars," explains one astronomer. "I am focused exclusively on quasi-stellar objects," says another. "Supernovas are my meat", "Comets are mine," "Solar interiors," "Stellar magnetics," etc, *ad infinitum*. Everyone has his specialty and studiously ignores anything outside it lest he be criticized for his temerity in poaching on some other astronomer's preserves.

So it goes on and on. Invariably each declaration is followed_by a halfhearted apology for expressing their indifference to the discovery, "It's not in my field, and I prefer to say nothing until the specialists in that have had their say," or words to that effect. Everyone is too afraid to speak out and express an opinion which has not been well vetted by his fellow travelers. The inevitable result is the sort of stagnation which leads to a requirement that researchers must explain *what they expect to find* before they receive any funding to look for it. By definition you must venture into unknown territory before you can find what is there. This is elementary logic. But sadly it is all too often ignored by the federal government and science bureaucracies as well as (worst of all) by the boobs in the federal Congress.

As recently as the 1960's physicists were still seeking to discover monopoles, presumed particles which possess only a north pole, or perhaps only a south pole. It was all to no avail, where it remains to this day. But most of the steam has gone out of their research and the topic has been dropped from the textbooks without any attempt at explanation.

We are still seeking gravitational waves, which we are certain must exist but have yet to be officially 'discovered' despite the odd reality that we routinely observe them without recognizing what they are.

In effect, with gravity waves we have a phenomenon we recognize what it is while with monopoles we have a phenomenon without explanation and no great discoveries of monopoles, magnetons, or even quasi-magnetons has been forthcoming. It would have been a nice touch, if only it had been successful but the world does not always work that way. So here we have largely unexplained phenomena which merit discussion in the textbooks but are usually to be found only in conversion tables, where they appears as a parent source for its relations to coulombs, farads, gilberts, cm/sec^2, etc. It is all delightfully circular and we are left wondering whether electromagnetism and gravity are merely figments of the imagination, or just what is going on? We can define them. We can measure them. We have given them names, but what the devil are they, and how do they work? These questions are still not answered. Perhaps those angels who have been busily dancing on the heads of pins could tell us… or perhaps not even them. They may turn out to be notional artifacts rather than real entities; an analog to the wavelength phenomenon. The only things we can be certain of are that magnetism exists and we can make use of it.

One other item is certain in this context; magnetism is a product of energy. It attracts/repels objects which are comprised of energy. We can produce magnetism by wrapping ordinary copper wire around a cardboard cylinder and passing an electric current through it. Hysteresis around unshielded high tension lines can play hob with radio and TV. Our Earth is enveloped in a magnetic shield which largely protects us from the harmful UV radiation of the sun. This more than suggests that we are not peering at some alien construct from another universe. Put all these little tidbits together and we may regard the relationship to ordinary energy as proved. But it still does not address the deeper problems.

We cannot answer all the questions about magnetism, but certain inferences may be drawn based on earlier sections of this cosmology. For one, it must be resident somewhere on the electromagnetic spectrum or it could not act upon other material occupants of the spectrum... such as iron, etc. Additionally, it has been proved to be a resident of our sun, of Earth and Jupiter, as well as other material objects. All this being the case it ought to be possible to make a few educated guesses and offer alternative possibilities about this mysterious phenomenon.

Over the years determined efforts to link gravity and magnetism have been dismal failures, but there may be some occasion to look at it from the standpoint of the individual phote. Is it possible that the collective gravity of dark matter also includes a magnetic capability if manipulated properly? The question is intriguing even as it may seem highly unlikely.

Or is it?

Occasionally obviousness becomes inconvenient. This ordinarily occurs when the obvious interferes with something we take for granted but otherwise happens to be wrong. Whenever this occurs the usual human response is to dismiss the obvious and persist in seeking another answer more to our liking and which conforms to our faith. So it may be useful to play a few games with magnets. Perhaps there we will be able to gain some fresh insight into the problem.

Take two bar magnets. As is the rule with magnets, one end of each bar will be the positive end, the other the negative end. Try to unite the two bars by joining like poles together and they obstinately refuse to cooperate, but put opposite poles together and they instantly fuse, so now you have a double-sized bar with opposing poles at either end. Oddly enough, however, the fused magnet does not have double-strength of the attractive power either to draw or repel a third bar. This argues that the 'magnetic' element is evenly diffused throughout the bar magnet, with x amount in magnetic capability per unit of volume. (I am seeking to avoid suggestive nomenclature here lest it may lead to unconscious conclusions through subconscious verbal traps, so please bear with me). Take this double-sized magnet and check to see if there is not some lingering magnetism at the midpoint... No luck there either

Try something else and see what happens.

Take an ordinary horseshoe magnet. One tip is positive. As expected, the other negative. Repeat the experiment and we get the same results. Now comes the critical observation. Bend the horseshoe magnet's tips so they connect and repeat the measurements. Nothing happens! The curved magnets have created a closed loop, or torus, and the paired magnets exhibit no external magnetism. This is always a part of elementary physics but the idea is strangely overlooked in astrophysics. Why this should be I do not know, but perhaps it is considered insufficiently arcane.

Some deeper probing of these toroidal magnets, carried out roughly half a century ago proved that while there was no external magnetic field there were domains of extreme magnetic activity going on at numerous points irregularly strewn about the interior; and this is interesting. It provokes the glimmer of an idea. It is not what we may call a clue. Nor is there is anything mystical about it, but it may point in the right direction even though it has no immediate relevancy to the question of the origin of matter in the universe we have indicated that our universe is toroidal. A closed horseshoe magnet is easily recognizable as a torus. Our universe is mainly so-called empty space interspersed by numerous galaxies. The toroidal magnet is strewn with tiny but highly active magnetic domains. Now we ask if it is possible to regard the magnetic domains in a toroidal magnet as the equivalent of galaxies in our universe? Or it this merely idle speculation on my part?

The only way to answer this is to try it on for size; and please bear in mind that arguing from a possible irrelevancy is quite different from arguing from a demonstrated fact. The first may be merely interesting. The second is truth.

Start by taking a closer look at the immediate environment of the growing torus which is our universe. In terms of our timekeeping it may be less than a millennia of years old and only as large as perhaps 600 light years in diameter, with a 200 light year eye at the center. But it has still not passed beyond its birth canal and so remains actively drafting ylem from the local space. Think of it as a titanic tornado passing through the atmosphere here on Earth. Its rotational orientation is by now well established, which means that the photes which comprise it are curved and basically indifferent to the ylem through which they are passing. Rather than being random in motion, their torques is roughly aligned and extended connected chains of photes permeate the nascent universe.

A notional artifact of these chains is an aggregate of magnetic fields as the phote chains required for our modern universe are crowded into a closely circumscribed volume of space. What would be the density of these photons? There are so many unknowns here that impossible to be certain of our estimates, but try cramming all the mass of the galaxies we can see into a volume of space which likely exceeded 400 cubic light years, but was fewer than 600 ly^3. It would have been a veritable pea soup of a magnetic field, running at least 10^{20} times greater than anything thus far achieved in normal space and probably around 10^{10} times greater than the maximum achievable by the European supercollider. It would be a persistent, though steadily declining flux density.

It would be within this electromagnetic soup than the initial particles and simplest elements formed. But this does not tell us how they formed. What processes had to take place for this to happen? How do we get from a pea soup of photes, admixed with a few electrons and other subatomic phote chains and come up with a universe pregnant with elements which can lead to complex molecules and even living organisms?

The clues are all there, but it remains to discover where they are and what they mean.

Go back a few years... to an era slightly before we started taking it for granted that whatever we believed was true was, in fact true. We had arrived at the concept of atoms and had more or less completed drawing up the periodic table so we could see how they fitted together. But we still didn't know how the atoms themselves were constructed. Along came a Britisher named Rutherford, who meticulously fitted various pieces of the puzzle together and... *arrived at a wrong answer!*.... *(It may be interesting to note that Rutherford was a student of Dr. Bickerton and wrote the preface to the latter's 1911 'The birth of worlds and systems', arguing that the sun consisted of anthracite coal. It illustrates the changes which have occurred during the century between the era when Bickerton's book was published and this text is being written. In other words, beware of over confidence in the state of our present knowledge.)*

At any rate, Rutherford looked at the layout of our solar system and concluded that atoms were constructed along the same lines, with a big nucleus corresponding to our Sun and circled by electrons corresponding to planets, all in nice, circular orbits *(what other shape could they possess since everyone knew that a pure circle is a perfect circle, and God only works along perfect lines? After all, the Pythagoreans and Aristotle said it so it had to be true. But then again, this remains a pretty sloppy universe, so who knows?)*

Rutherford's idea quickly caught on, and as recently as the 1930's high school textbooks routinely depicted the Rutherford atom as the last word in science, though physicists were openly discussing new ideas, among which lurked a 'fuzz-ball' atom where the electrons were regarded as tiny blobs which darted in and out from the nucleus. This concept, along with minor variants, occupies center stage today. As a demurrer, there is still a logical disconnect in here somewhere, so we must review the matter in more detail. We may accept either of the Bangs, Big or Little, or even the simple no bang tornado origin to kick-start the universe. We may also accept the work of the big collider boys and their thorny thicket of particles. Throw all of these into the pot together and shake it a bit. After a while it comes to a boil and we pour it out and, *voila!* We find hydrogen and helium, with perhaps a smidgeon of lithium thrown in for good measure. And after that we start getting stars as nuclear cook stoves busily spewing out myriads of heavier elements when they explode.

But how do we get from independent particles to the elements? Nowhere do we hear much about that little detail. It seems to be taken for granted that these elemental particles spotted their necessary counterparts wandering about inside the primal fireball, engaged in a flicker of flirtation before getting married, and then set about picking up a family of electrons or whatever in order to breed up to atoms like Iron or uranium, or something recognizable by today's audiences.

I do not really enjoy being sarcastic or snotty, but it would be interesting to see what might be forthcoming if some of our high-value talent were sent back to school in order to teach them how to think in an orderly fashion. That doesn't necessarily mean they will always be right. Logic is only as good as the available data upon which it is based, but deliberately eliding over logical gaps in an argument in order to mask them from consideration ought to be left to politicians, preachers and lawyers. It should never be regarded as a tool in the scientific arsenal.

Let us now examine the problem in a logical manner and, since the Big Bang idea has already been shown fatally flawed in simple geometrical terms (i.e. See Vlatavy's *Geometry of Einstein's Relativity Theory)* while the no bang model is unlikely, I shall restrict it to the little bang, low brisance, approach. It may be regarded as factual to argue that no new energy has been created since the birth of the universe. Energy is not something which is manufactured. It may be altered in its manifestations. It may be thought of in many different ways. But it is not created *ex nova*. Nor do we mean to imply that no additional ylem has been co-opted and absorbed during the toroid's passage through the dark matter which underlies our view of reality. Some may well have been acquired along the way, but it is also likely that there has been s seepage as elements of our universe fall back into the ylem and are lost forever. In terms of the percentage of initial mass of the universe both of the latter possibilities are probably negligibly significant and may be ignored pro tem.

Given this premise it follows that the electromagnetic force of the emergent universe, expressed in gauss, must have been confined in a relatively tiny region, perhaps somewhere between 125^6 and 125^{15} cubic light years, or between 1% and 100% the size of our galaxy, over a time span ranging between 1,000 and 500,000 years.

When we consider the myriad galaxies of our universe being crammed into that primal space this seems reasonable enough. Next add to that the stray magnetism of the intergalactic regions as well as the random interstellar flux. The magnetic flux density within the emerging universe would be trillions of times greater than anything man has ever achieved, even in his giant colliders... and it would be sustained over a vastly longer period of time.

The difference between the product of the colliders and that emerging from the torus stems from the insulation of colliders from the influence of the toroidal structure of the universe. Colliders smash the target atoms without regard for the handedness of the universe, hence we obtain evidences of the six-universe geometry as occasional, short-lived by-products of the six real or potential universes of our family.

Now let us visualize the condition of fragmentary phote chains caught up in this eruption. Many chains would be torn asunder while even more new ones would be created. It would be a vast, chaotically churning mass, with occasional streaks of light, both visible and below/beyond that, into UV, IR, X-ray, and even deeper ranges, racing around through the torus. The solitary hint of order would be the curve imparted to each phote when it was co-opted as an element of the torus.

The human mind has difficulty contemplating absolute chaos. It seeks to reduce matters by sequencing them, by compartmentalizing events so they seem rational when in truth there is no rationality. Those conditions apply here. Dozens or hundreds of things are happening with neither rhyme nor rhythm. Longer phote chains are created, then snapped and randomly fragmented. Chains acquire new photes, then lose them almost immediately. A solitary phote might be joined by 50 others, then break down into a dozen or so solitary photes plus several fragments running anywhere from two to a half dozen. Additional to all this would be a flux of neutrinos which had been manufactured from incidental pairings of photes formed in side-by-side unions where the opposing poles created a neutral alignment with no convenient 'snapping' point to facilitate separation; implying extreme smallness, zero polarity and a next to minimal mass equivalent to two photes….which would give them a ½ spin.

An inevitable concomitant to all this would be a tendency for multiple simultaneous warpage of the longer phote chains so they bend back upon themselves to form a closed loop, with the head firmly contacting the tail, as it were. At this point we have solid particles, i.e., electrons and an admixture of protons and other stable subatomic particles being formed. As a note here, while we may be skeptical of the reported manufacture of electrons in magnetic fields above some 50,000 gauss the skepticism is consequent to a lack of confirming data, but the logic is consistent, i.e. if 50,000 gauss cam manufacture electrons, what happens in gauss levels of 1,000,000 or more? Do we get protons? Are hydrogen and helium produced? Theory is silent over the prospect but the concept is encouraging.

So I may be egregiously wrong. But I may also be perfectly correct. Learning which is true is a task for the future. We cannot argue the validity of a theory for which there is no evidence, but it is consistent with the goal of constructing a framework capable of pointing in potentially useful directions and fitting new discoveries into a coherent pattern. Which is what this book is all about. But my argument is directed to the few who possess the courage to address the problem with open minds.

Either way, if correct this would not be the extent of the magnetodynamic action in creating these fundamental particles. Visualize an ordinary rubber band. This would roughly correspond to a phote chain which has recurved back upon itself to create a closed loop. Now take this rubber band and twist it until it is reduced to a tight knot, or wad. At this point a phote chain would form a stable configuration with negligible internal stress and few if any regions vulnerable to disruption by outside forces.

The constituent photes would remain intact, but their magnetic/ charge/mass qualities will be tracking endlessly around the particular strand which is their frame.

If this sounds rather esoteric it probably is. In an odd way we might think of it as a tiny, endlessly tracking carnival roller-coaster with its passenger cars in perpetual motion, but this would be a remote analogy. I am talking of something incredibly small, possessing a virtually invisible charge, mass and attraction while being capable of elastic extension and contraction.

I would further expect that the charge/mass/gravity feature would be generally inactive and distributed evenly throughout the phote and the focus effect would only manifest itself when it was impinged upon by an outside force.

As an added note, the more complex the atom the more tracks will exist in any given atom and the greater the complexity of their orbits. But the raceways for any specific atom's electrons meat pass through the nucleus and thus interact a few million times per second with similarly captive electrons. If the complexity of the orbits is too high there will be occasional conflictions. For the most part these will result in momentary ionizations but if the confliction is too severe it can result in fission. This is merely a random hypothesis which may well be in error, but either way it does not affect the logic of the remainder of the argument. Call it a *post scriptum* and be done with it.

With this I end our journey through the origins of the megaverse and universe. It is not really complete and there is more to be said, but next we need to turn our attention to various aspects of the growth and development of our particular little universe... the one we see through our telescopes and can detect with our instruments.

The focus will concentrate on the places where the model advanced thus far requires adjustments in existing theory. Where little or no modifications are required there is scant reason to reiterate established conclusions.

XIII

WHOLLY HOLEY OR WHOLLY HOLY?

To judge by the complacent smiles on the faces of practitioners of the arcane art of cosmic genesis we may well feel ourselves constrained to approach the question of galactic origins and the evolution of stars with contrite hearts and prayers addressed to the purveyors of Holy Writ. They have the subject wholly in hand, as the multitude of erudite texts attest.

Despite all the self-inflation there are a number of interesting little anomalies in current modeling of the evolution of galaxies and the course of stellar evolution within those galaxies. Something is wrong, but no one seems aware of it... or if aware, they studiously ignore it because existing models neglect to provide us with any ideas about how to account for it. All of which makes it fair game for me to chew on for a while.

Firstly, the Hubble constant does not allow enough time to get from the supposed era of the Big Bang to the present. Secondly, the models fail to explain how the first generation of stars got started, how they were distributed within the nascent galaxies as well as the sequence of generations in stellar evolution.

That should do for starters. After that, things start to get complicated!

An abbreviated recapitulation of existing dogma illustrates the problem. First we have a 'Big Bang'. Existing dogma illustrates the problem. Erupting with enormous fury from a point approximately the size of a robin's egg it spewed out all the energy which comprises our universe in a detonation which peaked in the first few seconds of our human concept of time. At the start this energy was concentrated in a single hyper-particle which then immediately devolved into a veritable Niagara Falls cascade of progressively smaller particles before settling into stable configurations which we denominate as neutrinos, electrons, protons, and a few other candidates of greater or lesser significance; including an abundance of light in the form of an electromagnetic spectrum... which seems to be regarded as a separate phenomenon in this theory.

This is the start of things. Now comes an indefinite period of time to allow matters to sort themselves out. Somewhere in this timeframe the electrons sought out protons {or *vice versa, or perhaps it was a mutual seeking out*} and promptly set up joint housekeeping so we arrive at hydrogen. How or why this happened has not really been explored; just taken for granted.

The newly manufactured hydrogen gradually formed into clumps and organized communities, thus giving birth to galaxies. But before the galaxies took form this hydrogen broke into suburban communities and commenced fusing into focused droplets which, once their internal gravitation kicked in strongly enough, coalesced into titanic stars dwarfing any we see in the heavens today.

Frictional pressure from the self-gravitation heated the infalling hydrogen gas to a point where nuclear fusion occurred so the explosive energy of the newly manufactured H-bomb counterbalanced the gravitational pressure to create a gravitational equilibrium. This nuclear fusion resulted in the production of helium while spallation probably led to the manufacture of lithium, and possibly odds and ends of carbon. Etc. These stars were metal poor as well as so massive that they burned themselves out in a few thousand years, then collapsed gravitationally and burst forth as supernovas, expelling enormous quantities of lighter elements into space before this first generation star subsided into one of three configurations, a white dwarf, a neutron star or a black hole, with the final destination depending on the residual mass left over after much of its mass was cast off in the explosion.

The next generation arrived when enough of this ejected material collected in a given region to start the process all over again. Implicit but not necessarily expressed in this is the premise that the second generation of stars will consist of bodies with smaller physical dimensions than the first generation stars. This would most likely be a by-product of the increased concentration of the mass of heavier atoms contained in the mix.

After this comes the manufacture of even heavier elements and the routines pretty adequately spelled out by existing theory.

But what about the start of things? Can anyone spot a few holes in the reasoning leading to the existing models? Suppose we take a closer look and see what is being said.

In the beginning the universe consisted of nothing but void. Then a tiny little fragment of that void got bored with non-existence and went Bang! Kapow! and regurgitated itself into the void with almost unbelievable violence, From this minuscule tear in the nothingness spewed out all of the energy which makes up our present universe. No mechanism capable of prompting such an explosive eruption is even hinted at so the best the theories can do is inform us that it happened so shut up and accept it.

Ok. We accept it.... Sort of; with the caveat that implicit in this is a return to the 'fabric of space' idea where the universe spews out from a rupture in some sort of containment system which kicked it in the seat of its pants in a mammoth elimination to end all eliminations. But still, this is what they tell us so we go with the flow. Accordingly, we move on to take the next step. Theory says the initial eruption involved a hyper-particle which emerged from the tear in space only to commence its devolution in something like a billionth of a second, initiating a complex series of fragmentations which culminated in neutrinos, electrons, protons and perhaps some massless photons which successfully wriggled their way out along with the other stuff.

A minor problem here arises from a realization that the available time frame is inadequate to account for the requirements presumed needed. So what to do?

The answer is simple. Postulate supraluminal speeds at the start of things! In other words, it goes faster than infinite speed before deciding to slow down to merely infinite speed somewhere along the line!

Next thing you know this budding universe breaks out of the emergent cloud to start breeding galaxies. Why? What motivated these clouds to form? No idea; they just did, so just accept and don't ask questions.

Somewhere in this mix we learn that the subatomic particles got together and had a risqué little house party. It must have been a pretty lively affair because most of these protons and electrons were born as a consequence. Many of the couplings led to hasty marriages in the form of hydrogen, etc; but a few pairings refused to settle down and set up housekeeping and only formed a brief pairing to manufacture neutrons; which split as soon as they got bored with one another.   Sarcasm   aside, here again there is no hint of a mechanism to account for these minor details. It is another case of being told to shut up and accept the sublime wisdom and erudition of the science clergy.

This ought to do for starters, but there is a bit more to be said. Take the fabric of space thing. The model I advanced earlier makes a 'fabric' of space wholly unnecessary, though it does call for a space which is densely occupied by dark energy. This obviates the whole idea of a 'rupture' of 'fabric' or any need for postulating a transcendental something lurking on the other side of space. It still doesn't tell us where space itself came from nor of the origins of this dark energy, which may be presumed to be ordinary ylem. But even if we reject the equation of dark matter with ylem it still wipes out several layers of useless complexity and defines boundaries.

The causes of the detonation which created the universe need little explanation. After all, most of us are acquainted with the mechanics of Atomic bombs; i.e. concentrate enough of the right stuff together in one place and it goes 'Boom!' without further ado. The degrees of freedom thing, with its six lineal dimensions is straightforward enough to require no comment and accounts for the handedness of our universe as well as the 'colors' of neutrinos and the multiple classes of subatomic particles which are distinguished only by their handedness.   All of this is to the good, but when we turn our attentions to the particular event which manufactured our universe there are more problems.

The first involves the duration of the of the birthing process, which may be thought of as the passage through the birth canal and represents the course of events which began at the instant of the eruption and progressed through the acquisition of successive concentrations of ylem until the density of the local medium and formation of the torus reached a point where we can assert that we now have a completed and independent smoke ring in existence.

Important things were happening during this interval and while we cannot quantify them we can at least understand them without recourse to some mystic kabbalah meant to bewilder us into believing the authors know what they are talking about.

In short, where current theory calls for an instant birth of the universe the model proposed here allows for a lengthy interval between conception and birth. And here I confess my inability to see why this concept is so far beyond the conceptual powers of the academics. I suspect too many of them have imbibed too uncritically of the biblical phrase where "God said, 'Let there be light, and there was light'"

Next problem: Take an ordinary child's balloon and inflate it until it is close to bursting; then stick a pin in it. Whoosh!, the balloon bursts and the air explodes out. The pressure release cools the outpouring air and it dissipates into the surrounding environment. Translate this into terms of a 10^{100} or so gigaton H-bomb erupting through the fabric of space and you will get cooling. Not a great deal to be sure, but the liberated slurry of particles spewing out of confinement are not apt to start congregating into new confinement of any sort, yet Lemaitre and his third and fourth generation acolytes of the 'Big

Bang' hypothesis require precisely this sort of reaction.

On the face of it appears we should expect the same of the model being advanced here. But here there are a couple of intermediate steps involved. First is the time scale. The Lemaitre model is an ultra-high brisance affair which goes to completion in a few brief seconds. The model we are dealing with may require several million years of time and motion, plus ylem acquisition before it finally emerges as a new born baby.

The second factor, which in a crude sense, may be regarded as the midwife in the process, is the accompanying electromagnetic field integral to the whole affair. The question is, can this be an acceptable accounting mechanism as a collection agent for attracting and then confining the ylem (perhaps admixed with hydrogen and occasional helium atoms incidentally manufactured at various times along the way)?

To answer this question we may turn to the enormously concentrated magnetic fields discovered in intergalactic space.

These are highly interesting, and certainly unexpected, but if correct they bid fair to solve any number of nagging little details. They appear to have been confirmed although I am unaware of the protocols. But they mesh so well with the line of thought I have been developing that I am strongly inclined to believe they will ultimately be confirmed and treat them as such even if the necessary proof is still lacking.

In the late 1800's the French mathematician Lagrange pointed to the existence of certain equilibrium points in the restricted three-body problem. For those unacquainted with the details,

220

the equilibrium points are regions where opposing gravitational forces cancel each other out so it is impossible to determine in advance of the event which way the target object will escape. Lagrange itemized five such points, of which three were unstable and thus purely temporary. Collectively these three are known as the straight line solutions. But two of these equilibrium points, known today as the "L4" and "L5" points are quite stable.

A few decades' later astronomers were startled to discover a number of small asteroids (the smallest about 60 km in diameter if my memory serves right) were occupying both the lead and trailing points of Jupiter's orbit. By now the known number within these two points exceeds a dozen. They were named 'Trojans' after Homer's account.

Getting back to the electromagnetic clouds in space, the remoteness of these electromagnetic galaxies (coining a name) prompts a suspicion they will eventually be recognized as occupying null-g or possibly null-e.m.f. spots between galaxies. If so, then they may be taken as either residual material left over from the little bang or, just conceivably, may denote places where the magnetic clouds are slowly accreting and may one day evolve into galaxies in their own right... or more likely, will coalesce to create globular clusters akin to those we find in our own galaxy. If the latter, then the parentage of these clusters, so long in question, is resolved.

As something of an aside here, I must point out that the spicule electromagnetic flares ejected from the poles of a few galaxies may wind up as seed matter for these electromagnetic galaxies. If so, the force of my argument is diminished, though not eliminated entirely. This is one of those situations where multiple factors may lead to a common ending.

Even if no connection with equilibrium points is discovered, the influence of electromagnetic forces in the emergence of galaxies and stars cannot continue being ignored. I acknowledge that Alfven's premature attempt to define the solar system almost exclusively in terms of electromagnetism was so flawed as to discredit any further efforts in that direction, but we are far enough removed from it in time --- and knowledge --- that we ought to be able to view other applications dispassionately.

As a sidebar to the role of electromagnetism in the universe, it is entirely logical for us to assume that it plays a crucial part in the concentration of ylem in the ancestry of our own universe. In a dense ylem slurry extending over an infinite space turbulence is inevitable. While individual photes possess a minuscule magnetism, it is far too feeble to attract other photes. But a large enough aggregate of photes contained within a turbulent cell of the right size, can exercise a distinct effect in drafting other free-drifting photes into itself while simultaneously growing in physical dimensions and increasing in density. Carried to an extreme it will arrive at criticality and explode.

No mystical fabric of space is necessary, and the detonation will be of low brisance since it begins early in its growth and never goes beyond criticality before exploding. Therefore it does not involve the manufacture of hyper-particles.

Unstressed but clearly implied here is the realization that our universe remains in the ylem as a casual smoke ring so there is no theoretical need for mumbo-jumbo about other spaces, though there is a clear distinction between objects within the torus and those outside the torus. Outside the torus space geometry is flat; within the torus it is curved. Inside a torus photes, particles and matter all generally follow a handedness appropriate to that imposed by their particular universe, exceptions being neutrinos and particles manufactured in colliders as well as a potentially random anti-particles which may be manufactured in black holes, etc. But even here we should bear in mind that just as the atmosphere through which the smoke ring moves is fundamentally independent of the ring and obeys its own geometry, the same would be true of the underlying ylem photes. So we would have two distinct geometries occupying the same space without in any way implying a paradox or a violation of some new and unexpected laws of the universe.

It is now time for a recap of the scenario to put it in a more orderly sequence and see what it looks like as an organized depiction of our newly born universe.

Forget 'fabric of space' altogether. It is merely a *deja vu* of phlogiston; a toy concept aimed at concealing ignorance useful in only as a pretense in order to make logical in idea what is inherently illogical.

We have one space only--- ylem space. Ylem is a specific name for the contents of the otherwise empty space. It consists essentially of discrete photes of light which is the tiniest unit of energy, characterized by a lineal geometry, and an electromagnetic/mass/gravity union contained within a space possibly as small as 10^{-60} cm^3.

At this point the three basic forces---electromagnetism, mass and gravity---are indistinguishable from one another so the individual

223

phote becomes the seed stuff for everything we know or is knowable. There may be occasional linkages of two or more photes in a manner similar to two magnets linked to one another *en chain,* or occasional neutrinos with photes linked together side by side with opposing poles touching. There may even be a few atoms of matter relict of earlier universes, though this is doubtful. Nature is messy enough that occasional exceptions may occur just about anywhere, so relict particles of earlier universes may be thinly strewn about in the ylem..

This ylem space contains at least one family of universes consisting of our universe, an anti-universe, and up to two additional opposing pairs of right angle universes, which collectively occupy the six degrees of freedom available to any mechanically oriented system. It is entirely possible, likely even, that there are families of universes similar to our own, strewn through the ylem, past, present and future. We can never detect any evidence of others nor of our own family of universes, (with a solitary exception of a direct collision of our particular universe with an errant universe born of a separate universal birth in another region of the ylem; which would be distinctly unlikely.

As concluded in the first section, >c can never equal C, i.e. the speed of light does not equal infinite speed, as Cantor defined it. The fact that each phote possesses a hint of mass precludes this possibility. Nonetheless, 297,000 km/sec is a quite respectable speed, so when two photes collide they ricochet away from one another at nearly the speed of light; exceptions being when an *en chain* hook-up or a parallel union occurs.

In the foregoing it must be born in mind that the individual phote cannot possess an external 'field.' If such a field were postulated then we would be required to stipulate an energy source which exists apart from the parent phote, and by supposing this we violate our thesis which calls the phote the absolute minimum unit of energy

A third, and highly unlikely exception, would consist of a few stray electrons or protons of any of the six degrees of freedom and relict of an earlier generation of universes. Apart from these few exceptions, the ylem contains an infinite number of photes tumbling about in a turbulent broth of aimless activity.

The general rules of turbulence in open space will apply here, so various regions within the ylem will have different ylem densities, which are generally ephemeral and continually changing. But this is not automatically correct. Taken as a single unit the electromagnetic mass/gravitational property of each phote is so minute as to be undetectable. Trying to deal directly with a phote to ascertain its e.m.f. would be roughly equivalent of seeking to detect and read the date on a dime dropped face down on the moon as seen by the Hubble telescope!

In *short, it can't be done Mein Herr! But it can reasonably be inferred if the problem is turned over to the bean counters. There they hit their stride—striving manfully to ascertain the next decimal place.*

The decisive final element which goes into the manufacture of a new universe would be electromagnetic confinement. We know this is possible because we are already doing it on a small scale, using electromagnetic bottles to confine antimatter in laboratories. I do not know what might be happening elsewhere around the world. But like it or not, the possible usefulness of antimatter as a military explosive dwarfing ordinary H-bombs is too tempting not to be an object of top secret military research in a dozen or more countries. But like it or not we may rest assured that it is happening. The solitary potential objection to the supposition that electromagnetic confinement plays a major role in the manufacture of universes lies not in whether it works or not but in whether it may be capable of self-confinement without the use of outside machinery. This is a worthwhile objection, but it is unclear whether it is substantive enough to warrant outright rejection.

I think not, but lacking intimate acquaintance with the subject I can only point out that it is probably a major component in the creation of universes and use it as an underlying factor in creation.

So the explosion happened and the ylem found itself pregnant with either two or six universes, all bursting with eagerness to be born. But this would not happen soon.

It might be thought of as a fetus in the womb. The process has a number of analogies to the birthing process among humans---though this is a dangerous equation and it is too easy to assign some mystical significance to the analogy..

To pursue the analogy we may think of the concentrated ylem cell as an unfertilized ovum with magnetic confinement created by chains of linked photes as the sperm which initiates the conception. This is followed by a fetal growth lasting an indefinite time and depends entirely on the successful retention of external ylem, i.e., if there is a plentiful supply of ylem to feed the growth is obviously requires less time than would be needed if it passes into a region of scanty ylem density. In brief, the gestation period may vary anywhere between a few thousand of our years and as much as a few billion years, so any effort to quantify the time frame will be mere wishful thinking rather than meaningful dogma. `Either way we choose to look at it, and regardless of the validity of my electromagnetic hypothesis as a factor in the gestation, ultimately criticality is reached and the nascent universe erupts from the womb and enters the birth canal.

The 'Big Bang' hypothesis has the universe erupting into existence pretty much as the ancient Greeks imagined for the birth of Athena; fully born from the brow of Zeus and instantly filled with galaxies An analogy appropriate to this hypotheses might be as a Fourth of July pinwheel tacked to a wall and spitting our galaxies in every which direction until it finally runs out of fuel, leaving any residue to slink dejectedly back into the megaverse from which it emerged.

The time factor involved in our hypothesis eliminates the mysticism implicit in conventional models and returns it to the status of a comprehensible physical process behaving along conventional lines. But it also eliminates all prospect of juggling matters to develop a handy calendar for pinpointing dates. The disappointment attending this failure will be bitter gall in the mouths of theologians and kabbalistic nerds of all stripes. The lust for precision can become overwhelming and a universe predictable as clockwork is the goal of every bean counter. One can only commiserate with their disappointment --- Better luck next time maybe. Returning to this universe, once the smoke rings are completely formed and are no longer drafting significant quantities of the ylem through which they are passing, we may regard the birth of our universe as a completed operation. It still needs to develop galaxies and stars and remains merely an inchoate smoke ring adrift in the emptiness of the ylem, but it has a fresh new geometry, a curved doughnut-shaped affair consonant to the shape of a torus, distinct from the flat, lineal geometry of the ylem, and providing confinement for the light being generated within the torus. The problem now is to transition from the cloud into an array of galaxies and stars.

This is fairly straightforward. By the time the smoke ring is fully developed it will have spent possibly several billion years in gestation and birth. It has grown from a newly fertilized cloud having an initial diameter of probably something on the order of 1,000,000 light years into a smoke ring with a diameter of several billion years. During this interval the progressive concentration of the collapsing soup together with the steadily increasing electromagnetic concentration leads to the wholesale production of hydrogen, with possibly some helium mixed in. Within this smoke ring we may discern tens of trillions of discrete magnetic domains which are analogous to the magnetic domains discovered in toroidal magnets, but these will be only a few tens of thousands of kilometers in diameter. It is from this phenomenon that we extrapolated upward to magnetic containment in the fertilization of the universe-to-be so there is nothing to be wondered at here and once these domains are formed there will be nothing to be surprised at when we extrapolate further down to globular clusters and individual star systems (although, to Alfven's chagrin, it does not work when it comes to the manufacture of planetary systems. He had the right idea but his idea applies on a far larger scale.)

There is much more detail to be presented in the context of manufacturing universes, and more variability than will provide comfort to those interested in coming up with clockwork universes, but I am not particularly interested in remarking every possibility. For that, whole encyclopedias would scarcely suffice. What I am concerned with here is the construction of a framework; a framework sturdy enough to hang data and theories on it without being compelled to create arcane, and often clumsy or downright rickety bridges needed to sustain if they are to have any hope of fitting into our universe.

I hope I have succeeded in this, but realistically I know I must have made errors.

Whether any of them will prove fatal is another question. No one is more venomous than an 'expert' who has been called to task. I fully realize what I have done here, but it was necessary. There is an old farmer's joke where he takes a 2x4 bat and whacks his mule across the head, then explains to an aghast urbanite, "I know it ain't polite, but first I got to get his attention!" I have deliberately stepped on any number of hypersensitive academic toes in challenging them to depend less time on airy formulas and kabbalistic mysticism and start thinking logically…without the contemplation of their navels routine. .

Next we consider the formation of galaxies and the generation of stars, and in the process arrives at a more acceptable age of our universe. Later, as a brief coda to my audacious cosmology, as an offering to the gods of contemporary science by presenting them with a final target for their venom, I shall presume to arrive at a rather different end for our travails.

XIV
STAR LIGHT, STAR BRIGHT

Somehow or another the grand poobahs of astrophysics have contrived to depict the birth of our universe as a rather sedate, even stately, affair; something on the order of Wagner's procession of the gods into Valhalla. First there was this 'Big Bang' which tore through the fabric of space. Then things sort of settled down. The rift closed and repaired itself while, lo and behold, the galaxies winked into existence, with their star lights lighting the brand new universe like freckles on a redhead's face, and promptly set up housekeeping in their new homes, presumably without even having taken time to go on a honeymoon together.

It is a delightful picture, but nothing could be further from the truth!

When we reached the end of the birthing process the torus which is our universe had finally separated itself from the ylem. It was passing through it, but it was no longer recruiting new ylem in any significant amounts. A few specks here, a few more there, never in any noteworthy quantities might happen but would be all. And to balance this were a few losses as debris filtered back out of the torus and into the ylem. Now let us exercise our imaginations and visualize the size of this infant universe and the contents therein.

Obviously we cannot know the physical dimensions of the newborn universe, but we can come up with some educated guesses. For one, we must postulate that the torus contained all the energy of our present universe, which includes all the galaxies, plus all the 'free' light and the mass and gravity which that implies. This requirement is also implicit in the 'Big Bang' model so we may take it as fact…although there is a remote possibility that the rift in the fabric of space idea rooted in the 'Big Bang' may have remained open for considerable time after the rupture so reinforcement energy may have continued seeping through for some time. But if this did occur then the outracing detonation will show distinct signs of a layering of galaxies, which we have yet to see.

Only El Greco could have done justice to these skies. The human eyes and ears are ill fitted to the task of visualizing the wildly swirling masses of invisible clouds of particles which stretched dimly across as much as several millions of light years of murky vistas.

Hydrogen atoms and traces of other debris coursed about within the confines of the expanding universe along titanic, spiraling journeys around the torus. Immense bolts of lightning, some hundreds of times larger than our puny little solar system, crisscrossing one another, arched across whole light years of space, reverberating deafening roars of thunder if only there were ears to hear it.

The nearest thing to this that we humans experience is a towering bank of thunderheads looming over the horizon with whole sheets of lightning flitting from cloud to cloud as they strive to equalize electric potentials built up by the friction of clashing forces within each competing mass. If we increase the scope of our imagining we must understand that we are surrounded and engulfed by the storm, beset on every side and at the mercy of forces we can only dimly comprehend and are incapable of resisting. We must visualize all the mass and energy of our universe pinched together within a volume of space not much larger than a doughnut where our own galaxy lies at one surface and the Andromeda galaxy at the other; roughly two million l.y. distant at the other.

A few million light years may seem a lot, but the universe made visible by the Hubble telescope extends to eight or ten billion years. In its compressed state we would be pushed and pummeled, shocked and deafened without the remotest hope of relief.... And this might continue for several hundreds of millions of years, or even longer, before matters began to improve. And just how would they improve? What form with it take --- and why?

Start with the progressive rarefaction of the particle cloud. As the torus expands the mean particle density dwindles. But this is only a relatively minor factor. More important is the wholesale manufacture of hydrogen. We already know that electrons can be manufactured in strong magnetic fields; or at least this has been reported by some pretty reliable sources. This raises the question of what can come of magnetic fields a few hundred million times greater?

And what can emerge from fields of utter chaos and completely random electromagnetic surges?

The human mind instinctively recoils from disorder. The whole of science reduces to an obsessive compulsion to convert disorder into order. In recent years the mathematicians have gone so far as to evolve a new discipline---*chaos theory*--- in order to extract logic from illogic! To a degree they have succeeded. They have snipped off a few corners and have even uncovered occasional fragments of patterns in random processes. But the underlying chaos remains intact. Nor   is   this   is simply my opinion, it is provable by simple logic and has been verified on countless occasions.

How?

By attempting to produce chaos! Every computer program which ventures to produce random numbers fails because the program itself must ultimately be rational. "Do this, then do that, and that, and after that ---" is a supremely rational approach to the irrational, which is most assuredly an oxymoron.

Go a little further and consider *pi*. We can discern an unknown total of repeating sequences of varying sizes, but all are relatively small and incoherent... which is precisely what chaos does. If all these repetitive sequences were eliminated the result would not be utterly chaotic. It would merely be a contrived pseudo-chaos. In likewise, we may expect to uncover occasional places where snippets of pseudo-chaos are extractable from the actual chaos occurring within the torus. But we can never really come to grips with utter chaos.

Pause now to consider a few of these snippets of pseudo-chaos. But cut me a little slack here because I confess to being in over my head. It would be altogether too easy to foist off dummy equations here and there; something along the order of "$x*y^{3+z}=0$ when $x=$*the speed along the a axis, the coefficient of resistance to the toroidal forces impelling conformity to the torus's internal motion, y an indeterminate variable dependent on the value of z, which represents the unaccounted forces exerted by the internal dynamics of the forming torus.*

It sounds impressive as all get out, and more than one savant has foisted equations of this ilk onto an awed but applauding audience as a means of concealing his ignorance. But actually it doesn't tell us anything. It is meant to conceal our ignorance behind a smoke screen of words and mathematical symbols: all "filled with sound and fury, but signifying nothing," redolent of the medieval philosopher Peter Lombard, who was known as the *labyrinth* because of his habit of spending endless pages of argument going around in circles before concluding with a "therefore" which baffled everyone by its incoherency. Ironically, Lombard largely succeeded in convincing his audience that he was a profound thinker.

So here goes my best effort to convey a few of the forces at play within the newly formed torus.

Item number one: Even with the little bang approach the initial outward thrust must be enormous; something on the order of a trillion or so kilo-megaton-hydrogen bombs exploding continuously along the pathway as the forming toruses begin to emerge. This would argue for a formidable initial ejection speed; perhaps on the order of a few thousand km per second. But almost immediately countering forces would develop to slow the emerging burst.

I have already pointed out that the little bang approach apprehends that it is really a rapid succession of contagious explosions as the emergent smoke ring ignites reactions in the dense, electromagnetically confining regions surrounding the initial detonation. These ignitions are the result of concentrations of ylem photes accumulating as a bow wave in advance of the forming torus. These secondary bangs will be omnidirectional so there must be countervailing forces working to muffle some of the initial momentum of the primary burst. More energy must be devoted to the task of creating and then maintaining the torus and incorporating any recruited ylem photes into the forming structure.

Nor can the influence of the lightning bolts be ignored. Logically, they ought to span arcs of as many as several light years. This is extrapolating from the dimensions of solar flares, which are underlying the still chaotic ylem. But the material which forms the torus itself is not chaotic and is, in fact well structured... at least for anything as sloppy as a universe. It is purely theological mutilation of logic which makes it out to be Wagnerian in its pomp and ritual.

And, of course there is the electromagnetic field itself. When I speak of fields running into the quadrillions of megagauss range I am not indulging in the juvenile game of "If I've told you once I've told you a thousand, million, billion times!" If anything I am being radically conservative. Our little galaxy is estimated to consist of between one hundred fifty and two hundred billion stars... {some recent claims would make it into the trillions} and for every star there is an estimated galaxy somewhere in the universe! so all numbers will necessarily be huge, and here I am deliberately estimating conservatively.

But there are a few little oddments of empiric science which may ultimately prove of assistance. One of these dates back to a little less than the middle of the last century, when the idea of employing magnetic containers to confine nuclear plasmas, whose temperatures range around $10,000,000^0$ centigrade! You have to admit, this is getting up there pretty high. This confinement may be achieved by superconducting magnets; which work; sort of. At least it is enough to show us there may be something worthwhile to learn in those regions.

A second item of interest is perhaps more psychological than real. Many of us have noticed a distinct ozone scent to the air after a lightning bolt has struck. This may, and I stress the 'may', argue that the continuous lightning display which must accompany either big or little Bangs, has the potential to play a defining role in the transformation of raw photes into electrons and protons, and thus to hydrogen. If correct then we have an unexpected source for the manufacture of significant amounts of the lighter elements. It also frees us from the need to poke into the 'innards' of stars to account for it all.

On this note I have exhausted my logic in discussing the immediate post-partum travail of our universe and its emergence into an environment capable of developing into galaxies and stars.

The reality is, humanity stands like a collection of ant scientists seeking to comprehend the world from their tiny perspective. We have no concise way of estimating the true parameters of the forces at play here, and I am certain in my own mind that I have scarcely scratched the surface. But there is a flip side to this... to the best of my knowledge no one else has ever gone this far, so to that extent I am blazing a trail through a trackless wilderness which may one day be develop into a paved highway... but not yet. There is more to the tale.

Precisely how long it took may never be known, but from this inchoate cloud of roiling particles, photes, light elements and electromagnetic surges aimed in futile effort to equalize the charges induced by random frictions, a few larger cells began to coalesce. As they grew in size something new began to happen. The local gravity of each cell reached a level where it exceeded the local gravity of the torus. In this it was actively abetted by an accompanying electromagnetic field which helped confine the steadily accumulating mass of hydrogen and other gasses and plasmas. The cell commenced drafting its own hydrogen ocean while remaining in the ambient cloud.

Ever greater masses of hydrogen rained down on the core of the cell, compacting it to higher and higher densities. Frictional heating began to rise as the core temperatures grew ever more concentrated..

Pressure from the infalling gases generated more heat as millions of metric tons of hydrogen accelerated their fall onto the steadily growing mass, ultimately culminating in the initial sputtering of spontaneous nuclear fusion as the core mass of the first star of our universe began to glow... At the start it was merely a sullen red glow, which speedily intensified to yellow, then white, and finally blinding blue as the star's core strove mightily to counter the infalling pressure by its explosive outward thrust.

It successfully slowed matters for a while, but nothing could repel the ever increasing pressure of the falling gasses. Its hydrogen was fused into a gaseous array of helium, lithium, carbon, nitrogen and even oxygen. It was short-lived star since it had to fuse its hydrogen into helium at a furious rate in order to avert immediate collapse, and its final collapse into a hyper nova occurred a scanty 50,000 to 100,000 years after its birth. In terms of modern stellar lifetimes this was scarcely long enough to spit at, but it was a start. It was the first star to shine in the new universe. Others would follow in ragged procession for it is not given to nature to be so concise. To imagine such titans springing forth in a synchronized cosmic ballet is simply silly. Here and there, across mega-parsecs of the torus a new star would burst into existence, briefly beam brightly, illuminating its portion of the heaven-scape before flickering swiftly into detonation. It and its scattered company of like stars constitute the whole of the first generation; short, sweet and wondrously explosive, our own sun can look back on the era with smug complacently and say itself, "Ah! Those were the days! If only I had been born in their age I could claim a share in their glory!

XVI
GO LIGHTLY GOLIGHTLY

There is a strange glibness in our theories of galactic evolution. The universe goes 'pop', and now we have galaxies popping up all over the place. But logically there must have been stars before there were galaxies. This is obvious. For example, how can there be a galaxy if there are no stars in it? The hyperstars cited in the last chapter become a necessary prelude to the development of galaxies by providing a number of widely scattered central core hyperstars capable of drafting the ylem in quantities required not only to initiate the conditions needed for the manufacture of hydrogen but also sufficient quantities of the hydrogen to provide material assistance in provoking the appearance of an array of first generation stars.

How long those super-stars reigned in all their glory can never be known, but given the sloppiness of nature we may rest assured their appearance was pretty haphazard. Be this as it may, they lasted only a brief era, perhaps merely a few tens of millions of years and produced a scant billion or so stars, before they departed utterly from this universe.

Why they passed is already well known and clearly charted, but for the benefit of those not well versed in this area of physics, there is nothing esoteric about it. An explosion pushes outward. Gravity pulls inward. Once a proto-star commences its self-gravitation it begins to fall toward the center of mass. In the process of compacting itself it not only generates pressure all the way to the inner core, it also generates heat through pressure and friction. But heat expands while the collapsing atmosphere is working to contract the sphere. Let the pressure exceed a certain point and the frictional heat and pressure becomes too great for the atoms at the core to resist. They collapse from the weight to create what is known as a *'Fermi gas'* where the electron shells of the atoms collapse onto their nuclei. The effect is rather as if the planets of our solar system all collapsed onto the sun, i.e. 99.9 or so percent of the volume of the system is lost at the same instant the density is increased multiple times. Nor is this this is the end of the affair. I know this was pretty much covered in the last chapter. But there is more to be considered so the goal here has a different thrust.

Think about it for a moment. We are back at square one…virtually returned to the ylem, with only electrons and protons at work in the core. But here again, there is more to be known. The Fermi gas is still confined within a pressure container… a pressure cooker if you will. It cannot escape from the core and the pressure from the infalling gas keeps increasing.

But this continuously exploding fusion bomb is still confined to the core regions of the star where the pressure is greatest. This means that the hydrogen in the envelope still remains, though not as abundantly. As the core is transformed into a Fermi gas the envelope continues pressing inexorably down on the shrunken core. At first the core continues to react by increasing its fusion rate, thereby consuming the remaining core hydrogen at an ever increasing pace. At some point along the line the last of the core hydrogen is spent but the pressure only increases. Now the only energy resource left to the core is helium, which is speedily exhausted. Lithium, etc., follow in close order and provide even less energy. Until iron is the only survivor, and it can produce no energy to hold back the crushing gravitation. But all these elements continue to exert their pressure until the last of the core's nuclear transmutations burns itself off.

In a glorious flash of incandescence the gravitational energy which had been pushing down on the core is released and the envelope which had been pressing down on the fusion reactor which constituted the core is blown away into space as a nova, as a supernova, or as a hyper nova, depending on the mass of the exploding star!

Now comes the tricky question. Where did all this light come from?

With Einsteinian curved space gravity becomes merely a notional idea, similar to numbers or the cash values of trinkets and lacking in reality. But when a Fermi gas takes the solitary route still open, it initiates a chain reaction and immolates itself in the fiery furnace of a continuing explosive fusion where the remaining hydrogen atoms are fused together and transformed first into helium the helium into lithium and the rest of the lighter elements up to iron.

Why stop at iron? Here comes another interesting little tidbit of information. Fusing hydrogen yields the maximum possible energy output. Call it 100 and avoid messing around with abstruse equations. On the same scale helium provides around 98, lithium 95, etc. The decline persists until iron is reached, where attempting to fuse it yields 0 energy. Coming at it from the other direction gives insight into the reason for this. Starting with plutonium, fusing it requires enormous energy and the end product is minuscule, amounting to a few atoms as reward. But instead of fusing, it generates a respectable amount of energy when fissioned.

On a descending scale when fissioned plutonium produces; call it 50 units of energy since fission is less productive than fusion. Then comes uranium, thorium, down to iron. Which again comes up 0.

In brief, iron is the end of the line. It costs more energy to fuse it than you can recover, but also costs more energy to fission than you can recover! And this is a key factor in the evolution of stars and planets

Now we return to earliest years of the galaxies and the little 'big bangs' found is novae, supernovas and even hyperbolas.

The explosions drive back the oppressing gasses, holding them at bay. This is termed a thermonuclear equilibrium and the star may endure for as long as its fuel supply endures; which may work out tor tens of billions, perhaps even a hundred billion years in the case of smaller stars, where the hydrogen fuel does not have to spend itself so swiftly.

This balancing act is trickier than it sounds. Our sun is roughly a typical mid-range star and it is easy to think of it as being in complete equilibrium, but it is not; not really. It pulsates ever so slightly as first one force gains an edge only to fall back when the counter-force compensates for it.   Since   the gravity pushes down the core reactor must burn hotter to compensate. But in burning hotter it pushes the gasses away, thereby reducing the pressure so the core does not need to burn so fast and the temperature drops... allowing the gassy envelope to fall back onto the core, repeating the cycle over and over in seemingly endless yoyo effect.

With our sun the oscillation is minor, amounting to a pulse magnitude of barely 2 or 3 km either way but a massive star which is far advanced in burning off of its core hydrogen might be pulse amplitudes running to several tens of thousands of kilometers.

Ultimately the sun runs out of hydrogen, but even then all is not lost. The burning of hydrogen is achieved by fusing atoms of hydrogen together to create helium, with the release of energy as a by-product. So several atoms of hydrogen are lost to acquire one atom of helium plus considerable raw energy with which to blow the infalling gasses back a bit.

But I am ahead of myself here. So far as the super-stars of the first generation are concerned there can be no equilibrium point. No mere hydrogen fusion bomb can resist the relentless pressure of the infalling gasses. It would be like trying to light a candle under Niagara Falls!

The internal reaction burns ever faster and faster, hotter and hotter as it escalates the fusion sequence to burn not only hydrogen but the newly produced helium, and lithium, carbon, nitrogen oxygen, and every other element up to iron, where it reaches the limit for energy producing fusion processes. All this occurs within a few seconds as the last defences of the dying star collapse and the outside gasses descend with implacable ferocity.

But still, the dying star will have its revenge. The fury of the collapse is ultimately a failure. The gravitational energy of the collapse releases a torrent of energy which adds to the dregs of the star's thermonuclear reactions to manufacture a titanic hyperrnova which spews quantities of elements running all the way the periodic scale, extending past iron to hurl some transuranic elements the torus and enriching the heavens for the next generation of stars.

Now for a quick non-technical word about fusion/fission reactions. As a handy reference point, the hydrogen/hydrogen fusion produces the maximum free energy. Hydrogen/helium fusion provides a little less. Helium/helium even less, lithium less yet, and so on up the periodic chart until iron is reached; and there it ends. From hydrogen to iron fusion reactions provide a new plus of energy. Energy is exported from the reaction. After iron any fusion reactions require a net energy input and you run an ever increasing energy deficit each step of the way.

From iron the process actually reverses. To achieve a theoretically useful energy plus you have to get into the fission business. In brief, going up from hydrogen to iron you employ fusion techniques, but these are steadily decreasing. From iron up to plutonium or above it is a matter of fission, with the heavier the element the greater the energy production from fission, with iron being the zero point where fission or fusion are both energy neutral processes. It is interesting to note that the ancient myth-makers were in a sense correct when they decided that magic would not work in the presence of iron. Iron is the end of the line in either direction.

Such was the first generation of stars.

We have now reached a point where it is time-out for an assessment of time. Henceforth any references to the age of the universe will use the moment when the burgeoning torus definitively split away from the underlying ylem; which is not to say that it escaped the ylem. No matter where it strayed or how it grew there was always the underlying ylem. Just as a smoke ring cannot escape the air through which it is moving but continues nonetheless to be a discreet entity, so is the torus a *drang en sich,* a thing unto itself. Now I ask how old is the universe when the second generation of stars began cropping up? There is no way to be precise but we do know that the estimates made by relying on the Hubble Constant are invalid because they are premised on a flawed assumption. This means that any guess is almost certain to be a flawed guess, but it may be possible to establish a few parameters. So here goes.

It takes vast amounts of energy to move an entire universe, even if it is only a truncated version such as I have suggested here. The energy contained in the initial eruption might exert no more than a few trillion, 100 megaton H-bombs being detonated in a region possibly several light years in magnitude... which means that the unit of force per cubic kilometer was fairly low; though cumulatively entirely capable of accelerating hydrogen and helium gasses to reasonably acceptable speeds, even if we restrict our computations to 1/6 of the total in order to accommodate the six real and potential universes at birth.

Somewhere between 100 and 500 km/sec would seem a decent starting figure to suggest from this explosion, with a half billion years between the initial separation and the full grown torus. But there is really no rule which ordains the construction of the torus must start promptly upon the initial explosion. There may have been millions of years between explosions. It really makes little difference in the outcome, save perhaps to the bean counters.

Such matters are their bread and butter and I cheerfully leave that little chore to them.

This about wraps up the first generation stars, but before I go to the next generation it is a good idea to summarize some of the arcana of galactic and stellar nomenclature lest I lose some readers by discussing matters alien to their usual wont. The next chapter ought to clear away some of the debris

XVII
OH BE A FINE GIRL—KISS ME!

This is not an invite and I doubt seriously whether any young ladies could avoid gagging at the prospect of kissing a decrepit old fossil such as I, but those who are *au courant* with astronomy will instantly recognize it as the mnemonic identifying stars by their size and brightness. The tale begins back in the early 1800's, an era when a one meter telescope was a marvelous invention of modern science and the acme of technology was the *octant* (later to be refined into a sextant.

About the only way of distinguishing between stars was their brightness and color. But there was a problem with this. In the old Ptolemaic astronomy, where stars were all implanted in a crystalline sphere circling the Earth all stars were postulated to be at the same distance, ergo, measurement of brightness was a perfectly realistic method of distinguishing between them. But the Copernican revolution, along with Kepler, Galileo, *et al*, pretty well eliminated such conveniences. An intrinsically dim star in a nearby region of the galaxy might appear much brighter than a hot giant star which chanced to be much further away. But despite this objection it was realized that even a flawed way was better than no method at all; so several different approaches were tried. One approach employed the Greek alphabet to designate the comparative brightness of the stars withinindividual constellations. Thus we might find *Alpha Centaurus, Beta Centaurus, Gamma Centaurus, etc.* This was a pretty good approach, one which is in common use even today as a means of localizing and designating individual stars or other features within our galaxy.

Tossing another crumb into the mix, a fair number of stars are paired, or even tripled, as in the Centauri system, *Alpha Centauri* consists of three separate stars, and these may be designated *Alpha Centauri A, Alpha Centauri B and Alpha Centauri C,* with the initial Greek prefix always rendered in lower case letters and the modern letter suffixes in capitals.

Confusing you say? --- You ain't seen nothing yet! Just bear in mind that bean counters cannot endure the disorderly. For them to purr with contentment every 'i' must be dotted and every 'T' crossed before they can sleep soundly at night. 'Prissiness' is the first and last refuge of the classical bean counter and astronomers are usually capable of out-prissying the most stodgy CPA or IRS agent; which is no mean achievement.

The other approach was not so successful. Still burdened by a supposed necessity to make astronomy conform to the dictates of churchly dogma, most astronomers resolutely ignored reality and (perhaps) subconsciously went back to the crystalline shell model of the universe.

Starting with the a.b.c. standard European alphabet they began classifying stars by size, brightness and color. Once finished with this task (a truly monumental one at that) someone shuffled and dealt the cards, trying to assemble them into some sense of order. In the process they arrived at the present arrangement. Casting out most of the alphabet and assigning the erstwhile residents thereof numerical gradations within the letter categories, they arrived at the existing classes and subclasses. Starting with 'O' they become O,B,A,F,G,K,M,--- hence 'Oh be a fine girl, kiss me', where 'O' type stars are the brightest and hottest,

'B' stars are relatively cool but exceedingly large, hence bright, 'A' stars are generally smaller and dimmer than 'O's, etc. Predictably, our own star was assigned to the 'G' category, (after the Roman Gaia=Earth) where it starts at G_0 with smaller, cooler versions of much the same spectral type are assigned to G_1, G_2, G_3, etc.

The system is workable but cumbersome, which makes it eminently fitted to the astronomical predilection for obfuscation. Pray remember that as recently as 1800 astrology and astronomy tended to be embodied in the same persona and many state astronomers were expected to cast horoscopes, forecast weather, and generally fill the robes of King Arthur's Merlin. If the court astronomer failed to couch his language in adequately arcane syllables he was quickly discredited and a new astronomer-royal would be installed in his place... which could result in an empty stomach in the failed astronomer. Truly, old habits die hard! The real marvel is that so many of the astronomer\astrologers actually produced remarkably good work despite the handicaps imposed by their lack of equipment capable of acquiring needed information.

As recently as the dawn of the 20th Century, many reputable astronomers remained unconvinced that phenomena such as the Andromeda Galaxy were outside our own galaxy. Those fuzzy little objects which clogged up their telescopes were nothing other than gas clouds in space --- identical to the clouds of Earth. "Nebulas" they called them, and they were dismissed as bothersome nuisances which obstructed their views of the stars which lay beyond them.

As something of an apology, I have deliberately omitted names and dates in this little historical digression. It has been roughly 60 years since I read about them and I do not regard it as needful to go back and review my sources. I concede there may be a few errors, i.e. attributing events as occurring in the 19th Century when in fact they occurred in the 18th or early 20th, etc. I accept without protest that not all of the motivations attributed to the *dramatis personae* may have been less inane, but such matters are of purely academic significance and not worth the time and effort to track them down. So if there are errors here treat my apology as having been given in advance.

The foregoing two techniques for naming and locating stars are mainly academic interest. Both are still in use, but neither provides much insight into the inner mechanics of stars. Now we consider a classification which provides genuinely useful information. And oddly enough, it is a product of an error in our understanding of stars. Our tale begins with the refinement of our measuring techniques which was perfected, around the start of the last century. The size of the Earth's orbit around the Sun was pretty well established and as such, it was quickly adopted by the astronomical fraternity. The mean distance of Earth from the sun is roughly 150,000,000 km. It requires a trifle more than 182 days for us to travel a semicircle around the Sun from one extreme to the opposite extreme, at which time it will be about 300,000,000 km distant from its starting point on the other side of the solar orbit.

300 million km makes a pretty long base line which provides us with a neat little trigonometric opportunity. If we wish to know the exact distance to an inoffensive little star a few light years distant we can take a measurement of the angle of said star tonight, then repeat the same measurement of the

star six months later. Now we have the classical angle/side/angle condition which any first year trig student can use to calculate the precise distance. And better yet, we have two angle/side/angles to play with along with the results from our angle/side/angle measurement.

But there were problems with this. Even the finest instrumentation and calculation could not assure we had achieved the needed accuracy of position and orientation of the six months later Earth, when the sun stands directly between us. Nor is the angler/angle/angle approach any better. In reading this please bears in mind that I am writing in terms of late 1800 astronomy. Employing modern equipment would be far superior.

So how to solve the problem?

It was the simplest way imaginable. Rather than trying to use the Earth-Sun-Earth axis as the shortest side, we split the problem in half and calculated on the Earth-Sun axis as the shortest side of a right triangle, and then six months later rechecked the measurements the same way, only when the Earth would be looking at the target star from the other side of the sun.

This cross-checking at first proved reliable measuring differences as small as 0.5 second of arc and provided us with the first reasonably reliable distances to nearby stars. Since then successive refinements have stretched our determination of stellar distances to approximately 650 light years, though the calculations would be somewhat iffy if only the parallax technique is used. Ordinarily measurements are repeated and then cross-checked several times before they are accepted, and even then they are still merely reasonable approximations... as, for instance, a given star is declared to be 1,500 light years from us. Implicit is the codicil, give or take perhaps 100 years. At that distance no one really cares. It is purely a case of close enough for government purposes; and actually far better than Congress achieves when it starts calculating the revenue it will have available to dissipate.

Not unreasonably, this unit of measurement became known as the 'parsec', a slightly corrupted form of 'per second' of arc. To add a touch of confusion, the parsec was first defined as 206,265 'astronomical units' (the a.u. being defined as the mean distance the Earth from the Sun). But that proved inconvenient and 'parsec' was redefined in the other direction as being 3.256 light years of distance. All this winds up giving distance measurements of 1 a.u. to represent the mean distance of Earth from the Sun, 1 light-year, (1 l.y.) the distance light travels in one year, and 1 parsec (3.256 l.y.). You pay your money and you take your pick. If by now you are not thoroughly confused you should be. So much for this little digression. I would never have mentioned it save for the uncomfortable fact that even well educated, highly literate folks in other fields get uncomfortable when listening to professional obfuscators who delight in resorting to argot in order to awe their audiences into reverent silence.

I also had a sneakily ulterior motive in mind when I resorted to this seemingly pointless digression into the task of ascertaining distances in space. It relates to an accidental discovery which cropped up the course of sky watchers seeking ever more precise distances estimates to remote galaxies. And to understand this you need to have at least a nodding acquaintance with these measurements and how they were developed.

Until the end of World War II the largest astronomical telescopes had objective diameters running in the vicinity of 100 inches (roughly 3 meters). They were admirable for poking around in our own galaxy but not so great when it came to nearby galaxies, including the great galaxy in Andromeda, or even the two Magellanic Clouds. It would be nice if we knew how distant either of these from our own. If we possessed that knowledge we would have a stepping stone to determine their sizes and go on to estimate the distances to even remoter galaxies. We might even learn the age of our universe and perhaps even its size and occupancy. But this knowledge was mostly lacking because everything depended upon possession of a workable measuring tape.

Luckily, the star *delta Cephei* lays within the circle of stars whose distances could be determined through their parallaxes; and it is a 'peculiar' star whose brightness increases and decreases rhythmically, completing each cycle in two and a half days. It was, as one young lady star gazer with a poetic turn of mind put it in a news column devoted to astronomy, 'as if the star was coyly winking at us and daring us to penetrate her innermost secrets'. This is not an exact quote, but it has been more than half a century since I read her column so I hope to be forgiven if I am wrong in detail. The spirit is willing but the memory is rather weak.

Today we have a whole class of pulsating stars known as *Cepheid variables,* but for many years they were largely ignored as being merely peculiar stars and there were more glamorous things to think about. In general, we knew they were characterized by reasonably well defined pulse rates ranging from around 2.2 days to as much as 43.4 days and that there was an inverse correlation in their peak luminosities, with the longer the period the brighter the maximum. This changed once someone pointed out an unanticipated oddity about these variables.

To explain, astronomy, like women's fashions, leaps from fad to fad.

Let an unexpected observation occur and every telescope on the planet swiftly focuses on it and theories swoop and swirl like buzzards zeroing in on a carcass in hope of gleaning a personal tidbit of meat from it, so the observer can claim a share in the glory surrounding the discovery.

Delta Cephei is a case in point. And what was the cause for this abrupt emergence into fame and glory after many centuries of comparative obscurity? The answer was not long in coming; and therein lays a tale.

For years it was more or less believed that pulsating stars were simple cases where a giant planet such as our Jupiter was in a 2.5 [or perhaps 43.4 or some in between] day orbit around the star, eclipsing it regularly. Its luminosity curve was plotted and then matched against other pulsating variables, but this was only of marginal use until it was realized stars with common pulse periods had comparable absolute luminosities. Translated, this meant that a pulsating star with a 2.5 day period, while seeming only a quarter as bright as delta Cepheid had to be twice as distant! In other words, old Euclid was still in business and our modern science hadn't changed a thing!

Now we had a workable measuring stick for the universe, or so it seemed. Astronomers promptly zeroed in on the class, checking periodicies and luminosities and calculating distances. The realization that some quite bright variables were known to exist in Andromeda and the determination of their period/luminosities was greeted with exultation. At last, we had a reliable measuring stick.

It also led to some hasty revisions in our textbooks as well as considerable discomfort. Earlier estimates of that galaxy, premised on the assumption that it was of roughly the same size as our own galaxy, were at complete variance with distances obtained from the cepheids. According to these the Andromeda galaxy was scarcely half the size as ours and much closer than we believed. Despite this discrepancy the cepheid value was accepted because it was so reliably accurate while the earlier estimates were premised on shaky assumptions.

But then the 200 inch scope on Mt. Palomar came on line and things started getting confused all over again. In calculating the distance of Andromeda by way of the cepheids it turned out that our neighboring galaxy was approximately the same size of ours and much closer than we had believed possible; which, in turn meant that our home galaxy was truly a giant among galaxies. And if this were true then distance estimates derived from other techniques were badly flawed all up and down the line! One way or the other, something was terribly wrong.

So it was back to the drawing board and in due course we had our answer. The solution turned out to be remarkably simple and wholly unexpected.

I could meander about at length, pointing out the nuances of the cyclic periods of cepheids and the luminosity curves associated with those periods, but it would be idle on my part. The important thing was the fact we actually did have a wonderfully precise tape measure for the universe. We just hadn't applied it properly. We had been logical, but had overlooked an important clue in our quest for absolute certainty.

It turned out that there were not one but *two* classes of cepheids, and by the luck of the draw we had comingled them!

The limitations imposed by the 100 inch telescope had compelled us to peer at the brightest variables in Andromeda because we could distinguish them from the blurry mass of surrounding stars. But our local cepheids we were comparing them to were generally in the direction of Andromeda though not really part of it. Instead, they were outlying stars of our own galaxy.

When the 200 inch came on line we were able to see the other, fainter variable stars and could contrast them to brighter variables having the same period!

Attention now veered back to the Milky Way (the name of our own galaxy) where we quickly discovered that the same disparity applied here. There were bright variables with luminosities arrayed in accordance with their periods, but there were corresponding dim variables similarly arrayed! In brief, these stellar populations were titled Population I and Population II stars and they were quite different.

Robert M. Kraft, late of the Mt. Wilson and Palomar Observatories did it more succinctly (and better than I could hope to do) in the introduction to his chapter on the Absolute Magnitudes of Classical Cepheids. (University of Chicago Press series "Stars and Stellar Systems" vol. III (Astronomical Data) pp 421, sic:

"Over the years, the determination of extragalactic distances has been based on the assumption that cepheids of a given period were essentially alike, regardless of the stellar system in which they occurred. The first change in this outlook was forced by Baade's distinction between the cepheids of Population I and Population II. It gradually emerged that the long period cepheids of globular clusters were old stars of comparatively low mass [Arp, 1955], their spectroscopic properties were similar to those of the nearby peculiar cepheid *W Vir* (cf Wallerstein 1958). Thus the need for modification of the earlier assumption was very clear; it was replaced with the idea that the period luminosity law could still be used, provided that only stars of type I (Population I) stars were considered.

"Recent studies suggest, however, that the concept of "type I' and "type II" is a serious oversimplification. It is probably more useful to regard age and chemical composition as the basic parameters regarding stars and stellar systems."

There are clues to the mechanism underlying pulsating stars. They are nearing the exhaustion of the hydrogen in their core region, so the gravitational forces are gaining ground, compressing the star. The star resists the pressure by increasing its burn rate so it expands and grows brighter; which in turn moderates the heat production, thus permitting the gravitation to recompress the core.

But my concern here is not with the technical details underlying the variable star phenomenon. I am more concerned with the fact that the raw data, the integration, the interpretations and the conclusions were all logically coherent and without any need to introduce 'bugger' factors to sustain them. We had finally achieved the much longed for dream of all astronomers: We had a yardstick to measure distances even as far as the Andromeda Galaxy, and from there all the way out to the verge of our instrumental reach! This was heady stuff, and before long astronomers the world over were complacently informing the awed plebes of their newest findings and promising that the uttermost mysteries of the universe would soon be resolved.

All of which brings me to the reason for my chatter about purely mechanical matters in a text where I am primarily focused on concepts. The two types of stars are not randomly distributed and their spectra are markedly different. Type 1 stars are now known to be the product of spiral arms and are 'metal rich' while Type II stars are located in the galactic halo and are 'metal poor'. This is both interesting in itself and highly significant in providing us with a useful road map to universal history and its evolutionary timeline.

When a Type II star largely exhausts the supply of hydrogen fuel in its core the core reactor region begins to collapse under the gravitational pressure. The heat generated by the envelope's pressure builds to a point where the core helium (which is a waste product of hydrogen fusion) begins its own fusion reaction.

The energy produced by this reaction is markedly less than that produced in a hydrogen reaction so the star must burn at a markedly greater pace which means the helium must burn hotter and more rapidly in order to resist the pressure exerted by the envelope. Of course the process is considerably more complex than I have suggested here, but the concept should be clear; the debris left behind by burning helium, together with the incidental higher level debris manufactured during the hydrogen burning phase accumulates and is spread throughout the main bulk of the star. As the core struggles to battle the crushing self-gravitation it pushes back and the star responds by expanding. In the process of expanding the core cools down fractionally (Please recall how the air which drains from an inflated balloon emerges cooler than it was when confined.) The star core does the same thing. It cools and damps its burn rate so the heat production slows. This creates a seesaw effect where the star alternately expands and contracts. It pulses and we have a Cepheid variable.

When it first initiates this cycle the star is still burning residual hydrogen circulated in from the mantle and actual helium burning is sporadic so its pulse period is a long, drawn out affair. As it gets deeper into its cycle, where helium, lithium and even heavier elements are progressively fused into even higher elements, the pulse rate steps up, progressing from several months to a few weeks, to a single week, thence to a few days as the core struggles desperately against the crushing gravity and its inevitable fate.

Crash!!!

The surrender, when it comes, is practically instantaneous, requiring only a few seconds to completion. The atoms in the core and inner mantle are literally crushed by the enormous pressure and collapse into a Fermi gas, shedding 99% of their volume in a tiny fraction of a second! It is rather like an inflated child's balloon when pricked by a needle --- not an exact analogy, but close.

The effect on the outer mantle and atmosphere is precisely akin to the detonation of a titanic H-bomb, magnified a few million times. Radiation pressure takes brief command of the process and probably around 80% of the total mass of the star is blown away and ejected into space, with the remaining 20% subsiding into a sullen existence as a white dwarf star, a neutron star, or a so-called 'black hole' (otherwise known as a Shwarzschild singularity, or simply as a singularity, depending on the whim of the author).

Nor is the blowing away process a simple affair --- which ought to come as no surprise to anyone who has had the stick-to-it-iveness to read this far. Titanic forces are at play here too. Not as titanic as the forces which gave birth to our universe, but strong enough to enjoy eating our little Earth for dessert, or as an after dinner mint offered by a restaurant to top off a gallant repast. This ejected material is permeated by a myriad fusions, fissions, momentary chemical unions, ephemeral elements, (spectroscopic observations of supernovas reveal the presence of the element Californium, which has a half-life of only a few seconds), secondary internal explosions; all occurring almost simultaneously in a grand melee of explosive violence.

This is not simple imagining on my part. Our H-bomb test firings have documented precisely this behavior on a far smaller tapestry. An ordinary nova is quite capable of ruining a perfectly good solar system. A supernova may slag all life on systems 40 to 50 light years distant while a hyper-nova of the sort generated in the opening days of a universe might compass a globe 500 l.y. in diameter and extinguish all life within 2,000 light years!

Now that would be a real bang-up job!

Admittedly, these numbers appear daunting, but this is a mighty big universe and all of it was compressed into a forming smoke ring whose outer perimeter was most likely a million or so l.y. in diameter by the time the first hyper-giant stars made their appearance on the scene.

As might be expected these hyper-stars did not blink into existence simultaneously. Nature simply doesn't work that way. The interval between the first and last apparitions of the hyper-novae may easily have extended over a billion years, with one or two blowing up every century, shining for 35,000 – 50,000 years until the core hydrogen has burned and then blowing themselves apart.

In brief, my earlier 'ball-parked' estimate turns out to be entirely too short. It is difficult to imagine this birthing process as being any less than two billion years, and it may wind up being refined to an even longer period.

The debris spewed out by the first generation hyper-stars includes quantities of gaseous atoms such as hydrogen, helium, nitrogen and oxygen, in progressively smaller quantities as the atoms become more complex. These atoms were probably highly ionized, consisting of protons and neutrons, with electrons to be scavenged out of the ambient soup; however it is also likely that a tiny fraction emerged as completed atoms. But no matter; either way it came out the same at the south end, i.e. we would find ample quantities of the major gasses drifting free in space.

Additional would be quantities of solid atoms culminating in iron, which is the end point of routine fusion since it requires the same amount of energy to fuse it as it requires to fission it. All these products were flung through space by the initial hyper-novae generated by first generation stars. It is this debris which provided the seed matter for the next stellar generation.

There may be reason to postulate a second generation intervening between the hyper-novae of the first generation and the Type II stellar generation; call it a cluster of Type III stars consisting of what are today alluded to as 'globular clusters', but this is by no means certain and will be addressed later, so for the moment the Type II stars are the focus of our attentions.

The debris from which they are formed is metal poor. Iron is reasonably plentiful, but that is not saying much. The still free magnetic fields enter into the picture here and the iron atoms tend to clump together, ultimately forming boulders of iron in space, boulders which may grow to a point where they start drafting other elements to their surfaces. When enough hydrogen accumulates the compression generates a new fusion reaction and a new, but much smaller star is generated --- a Type II, metal poor star. Over the aeons the galaxies which characterize our universe have formed, presumably each centered about one of the hyper-stars. And here again I must emphasize the scale of our universe. The 'Milky Way', which is the official name of our galaxy, consists of 150 to 250 million stars. Moving at the speed of light it would require roughly 100,000 years to cross it at the waist and perhaps 175,000 years to move from pole to pole, and a mere 35,000 years to depart Earth and reach the core. In other words, we have a pretty big back yard and it takes some pretty big events to make any noteworthy differences. If you insist on thinking small, them match these sizes against a routine supernova and realize there is no comparison; the supernova is a mere firecracker when matched against the forces which generated our universe.

This brings me to the crux of my reasoning. The chief characteristic of Type II stars is the deficiency in heavier elements, i.e. those of iron or above. The typical Type I star possesses relatively large quantities of iron and higher elements. A secondary distinction between Type I and Type II stars is their nearly complete separation from one another. Type I stars form the spiral arms while Type II stars comprise the halo. Blink out the Type I stars and the galaxy appears as a perfectly normal globular galaxy. Blink out the Type II stars and we see a runty spiral barely 20 percent the size of the Milky Way.

We find occasional Type II stars thinly strewn in the spiral arms, i.e. Barnard's star, which is within our neighborhood but has a highly eccentric orbit around the center of the galaxy; an orbit which takes it far out into the galactic halo, but few, if any Type I stars are outside of the spiral structure. The inescapable conclusion is our galaxy began as a normal globular galaxy and the spiral arms are a rather later afterthought --- which makes them either a third or a fourth generation descendent of the developed torus. As an added complication, there is evidence that a brand new, possibly fifth generation of stars is in the works. But this is irrelevant here. The point is the Milky Way galaxy, along with the Andromeda galaxy and other spirals, actually consists of *two galaxies*, a typical globular galaxy with a later developing spiral galaxy superimposed upon it! Those who dogmatically insist that the Milky

Way is a spiral galaxy may be taxonomically correct but they are otherwise blind to the reality.

We have recently discovered the existence a thin scattering of these ultra-metal enriched, and patently youthful stars embedded in the older debris of long-defunct supernovae in both the spiral-armed and halo regions of the Milky Way. These stars, which often exhibit strong beryllium and rarer lines in their spectrum, betoken any associated planetary systems so rich in heavier elements that Earth-grown higher forms of life could not exist there. They are early representatives of a newly evolving fifth or later generation of stars; which likely signifies an even older age for our universe.

As a note of caution here, it has been claimed that no two snowflakes are identical. I am unsure of the accuracy of this observation. There are a lot of snowflakes and only a relatively few analyses of their structure. So instead of counting snowflakes please note that no two leaves of an oak tree are identical nor are the leaves of an elm tree, but any observant individual instantly distinguishes between oak and elm leaves. So two with stars; there are billions of stars but in all likelihood no two of them are identical. However, like oaks and elms, classes of stars, i.e., Type I and Type II, are immediately distinguishable.

Where does this leave us with respect to the age of the universe? Most of the pieces are in place but they need to be assembled properly. There will be gaps here and there, but the basic outline is clear. Start with the mechanics of a smoke ring drifting in the ylem. Visualize it from the perspective of a cigar smoker blowing a smoke ring or, if the prospect of cigar smoke is too repellant to think of, try an ordinary cardboard box instead. Fill it with pressurized smoke and then poke a small hole in one side of the box and tap it on the opposite side. Do this properly and you can get some beautifully formed smoke rings escaping through the hole.

If you now imagine yourself perched upon a molecule of this smoke ring busily studying the neighboring particles of smoke you would measure their rates of recession from you, calculate them back to a universal zero point and, *voila*! you have the age of the universe: Day One!

All your textbooks would proclaim this number and you could smile in smug contentment. Of course you would be wrong. You would be wrong on many points, but no one would know, the reason being that your conclusion was premised on a faulty philosophy which called for a Big Bang casting the universe out from a single point. The most amusing item here is, the number it would continue to have a certain small value, but it would be a different sort of value.

Start with the negatives; no one would ever be able to decide the precise instant the universe came into existence. The period of separation takes time. Do we count the beginning from the instant when the ylem reached its initial criticality or when the final explosion began the actual process of forming the torus? This might have taken several tens of millions of years; not much by universal reckoning but still a goodly figure.

Another starting point might be when the torus is fully formed and the first hyper-stars began flickering into existence. This is probably as good a starting point as any but it may well have come as much as a couple of billion years after the initial little bang. I personally prefer to regard the instant of the initial explosion as the start-up point but I cannot fault anyone for adopting one or another of the alternatives.

Given my starting point I have to consider the likelihood that the torus has a diameter ten or more thousand light years before the stars began to appear. Then there is a minor little matter of location, and here the only certainty is that the central hole of the doughnut Is not it. As with any smoke ring, its motion inevitably carries it away from its origin.

In short, the potential value of the Hubble constant is more subtle and less certain than existing dogma pretends.

Try measuring the recession speed of the individual molecules from the molecule you are standing on and you quickly realize you are dealing with a succession of growth spurts, and the Hubble constant cannot be a constant, but it may prove useful in ascertaining the ylem densities through which it has passed by measuring the growth spurts, the tiny anomalies in apparent recession speeds indicated by slightly conflicting speed determinations seen from differing regions of space.

To explain, consider tree-dating by means of its growth rings. A tree will add a growth ring for every year of its existence. With each growth ring the circumference of the tree trunk increases slightly. The width of each individual ring is determined by the amount of rainfall received during the growing season. A rainy year will exhibit a wide ring while a drought year will show only a narrow ring. Apply the same reasoning to the torus. Here the controlling factor is the density of the ylem through which it is passing. Add to this the fact that the torus is immense and is to a degree disconnected so various regions will be capable of expanding at a slightly different rate than others for a period of time while it is being passed through or is passing through a somewhat denser region of ylem. But it will be random so in another aeon a different region of the ylem will be affected so it largely evens out.

The differing measurements may, over time and with improved mechanical techniques, provide insight into the overall density of the ylem as well as an indication of its age and possible destiny of the torus as a whole.

It may also be of use in ascertaining the physical dimensions of the torus, though this is probably asking a little too much of old Hubble.

Taken as a whole, Hubble is not hopelessly obsolete after all. It just cannot be used for its original purpose.

Assembling our time line with any degree of accuracy is obviously an iffy proposition but the line itself is consistent. Take it step by step and see what happens.

Step I: Turbulence in the ylem implies density variations in the photes, with high and low concentrations throughout. When a high enough density is achieved the phote magnetism and gravity begins drafting additional ylem until criticality is reached and an explosion ensues.

Step II: This explosion is not a 'Big Bang' arising from a point source. Instead it is a series of 'area bangs' commencing with a single bang over a region possibly as large as our galaxy, but more likely around the size of a typical globular cluster such as we find within our galaxy.

Step III: The initial 'little bang' manufactures a shock wave which transmits itself through the surrounding ylem, compressing it to create a series of possibly lesser bangs in all directions in a sort of chain reaction of chain reactions.

Step IV: The chain reactions occurring along each of the cardinal directions soon conjoin with one another of their associated explosions to generate self-directed spearheads of semi-coherent energy which pierce through the centers of these explosive chain reactions, resulting in what may be described as a hyperbolic paraboloid of six sheets [expanding cones pointing off in all six directions].

These six conic projections each have a different degree of freedom which became its signature, one matter, two antimatter, and three, four, five, and six, which may be characterized as 'other matters'. The six degrees of freedom are still to be found emerging from atom cracking operations, so there is no reason to regard them as exotic.

It would be a serious error to act the rigid mathematician in regards to these six eruptive chains since their destinies hinge upon the ylem density through which they pass and thus the size and frequency of the individual sub-explosions. For example, think of our universe as number one. Number two, the antimatter universe, might be moving in the direction of extreme rarefaction of the ylem and thus develop into a micro-sized runt universe topping itself off at less than a tenth our universe. This would make no difference so far as we are concerned since no parity of numbers is implied; only parity of opportunity.

Number three might encounter a density concentration twice as that of our universe and develop into one twice the size of ours. Number four possibly ran into internal difficulty where secondary internal concentrations prevented the development of a torus and the whole affair aborted while numbers five and six developed more or less along the lines of numbers one and two.

In short, we should regard this phase as merely another example of the messiness of nature. Since we can never know the truth of the matter, I shall henceforth ignore the fates of the other five universes and all my arguments will center on our personal universe without trying to count the number of angels on the heads of pins. It is far simpler that way.

Step V: The initial stage in the creation of the universe was permeated by titanic frictional stresses leading to exchanges of lightning and other electrical displays within the individual clouds. Magnetic fields running into the billions of gauss caromed through the turbulent mix of primitive ylem and newly manufactured elementary particles, i.e., protons, electrons, neutrinos plus a smattering of more exotic particles. Included in the mix would be a few of the simpler atoms such as hydrogen, helium and perhaps occasional atoms even higher on the scale.

Had there been any ears to hear, the expanding cloud was permeated by the sullen rumble of colliding photes and particles, admixed with bolts of lightning and accompanying thunder spanning entire light years. Sheet lightning and St. Elmo's fires provided dim Illumination and cast a ghostly glow which dutifully followed elliptical tracks ordained by the circulation of the torus.

Tidal waves of electromagnetic force surged wildly throughout the nearly formed universe.

Almost lost amidst the grandeur of the spectacle would be momentary sparkles of light as newly created atoms of hydrogen or helium were manufactured, with most, though not all, being ephemeral and promptly annihilated within the chaos attending the event. But a small percent managed to survive, ultimately to become incorporated into stars.

Step VI: Somewhere along the process the chain of explosions began dwindling as the ylem concentrations through which they were drifting thinned out. The turbulence lessened and magnetic and gravitational pressures created localized minor clouds of mostly hydrogen gas.

Another digression follows. Forgive me for doing so since I have deliberately sought to avoid the cacophony of citations and footnotes which characterize scholarly essays nowadays. Most are meant to deflect criticism by persuading the reader to assume the author knows what he is talking about and has expert opinion to back him up. But here we have an exception. It is not an *'op cit'* note but a suggestion aimed at verifying the accuracy of my reasoning.

Wherever we look in space we discover electromagnetic fields; around the sun, along with the Earth, Mars, Jupiter, Saturn, Uranus, Neptune and even the moon. All are known to possess them [The moon's e.m.f. is exceedingly faint]. Current belief calls them creations of the of the sun, Earth, moon, etc, but according to this model it is the other way around and in many cases it is the electromagnetic/gravity field which enabled the sun, Earth and moon to form.

It is likely the sun, Earth, moon and other planets returned the favor by locking the e.m.f. into place so we routinely detect them, and it is certainly not a rehash of Alfven's model even if he did have a noteworthy glimmer of truth, making him an astronomical precursor of Wegener and Continental Drift.

The existence of dark energy in space, the galactic electromagnetic field, the spiral arms, and recently the discovery of vast fields in regions of space; all attest to the prevalence of e.m.f. in every nook and cranny of the universe. Concentrated domains of free standing e.m.f. occur in the neutral zones between established domains and thus become focal points for the creation of new matter, new stars, etc. Seek them out and see if I am not correct.

Returning from my latest digression, these 'minor clouds' were only minor in context with the torus. To us they were unbelievably vast since they consisted of much of the energy and makings of our present galaxy; most perhaps, but not all. Smaller condensations were attracted by the superior gravity and magnetic ambience of the larger so they tended to merge.

A cognate situation today would be the greater and lesser Magellanic clouds tributary to our galaxy and destined to merge with it in a few tens of millions of years. By a more remote analogy is the realization that the great galaxy of Andromeda is fated to merge with our own milky way in a few hundred million years. Not enough is known about the Abel galaxies to hazard more than a guess but it is likely they too will ultimately merge into a single multigalactic galaxy.

Step VII: And here comes another digression. But this one is more directly to the point. While the torus was forming, a secondary process was developing.

To explain; in Part 1 of this work I concluded that electromagnetism and mass/gravity were different aspects of the same phenomenon and were an attribute of each phote. Logic required it despite the fact that persistent efforts to equate the two forces had repeatedly failed. Still both are attractive forces, though magnetism is attractive only when it is acting upon certain metals, primarily iron but not lead, etc. For its part gravity attracts everything. Magnets manifest themselves as looping patterns in bar magnets and planets alike, arching around from 'north' pole to 'south' pole. Gravity is omnidirectional. Magnets offer lines of force. Gravity simply *is*. To all appearances there is no realistic connection between the two forces. The only thing they seemed to have in common was their unique ability to attract things, but this is a big point to have in common. Force pushes. It does not pull. What mechanism can pull by pushing you? That idea simply does not compute!

But appearances can be deceiving. It takes some getting used to think of a diamond as being merely compressed coal and even greater imagination to think of aircraft wings being merely processed coal, even though both are simple fact. It requires far greater imagination to recognize the relationship of a human being to a blackberry bush or poison ivy! But this too is true. We are all the products of DNA and RNA, and though the chains are differently arranged they are constructed of the identical building blocks and are all akin to one another. By failing to expand their logic and cast a wider net those researchers who had sought to equate gravity with magnetism revealed their paucity of imagination. Had they looked for alternative perspectives the problem might well have been resolved by the Frenchman, Coanda, in the early 1900's; in which event Einstein would no doubt have come to a much modified set of conclusions.

And what was Coanda's contribution to the problem of gravitation? Known to science today primarily for his studies of surface tensions in liquids, he remains virtually unknown and certainly neglected for his role as the inventor and pilot of the world's first jet aircraft in 1916, roughly a quarter century before jet aircraft started appearing in the skies!

The airplane made its debut as a war machine early in the great "War To End All Wars" and improvements in the technology were widespread and frantic. Where in 1914, airplane design was not much advanced over the box kite, by 1916 and the battles of the Somme and Verdun entire squadrons of two wing fighting machines armed with twin machine guns and bombers carrying a ton or more of bombs ranged the front. They were rapidly approaching stalemate in the air, largely created by the limitations of the aircraft engines, which were both heavy and inefficient.

Coanda had been working with hydraulics for several years and hydraulics also includes air as a liquid. He reasoned that it might be possible to employ air in a more efficient way and forthwith designed and built a jet engine. It was tiny by today's standards and could easily be carried by hand but it worked. He then submitted his ideas together with a larger handmade engine to the French military, and by a miracle of bureaucracy unknown to the western world today, he succeeded in obtaining an engineless Spad. Equipping it with an even larger but still handmade jet engine and filling a fuel tank with castor oil, he climbed aboard, adjusted his goggles and pumped fresh air into the intake as he cranked the engine to life.The engine coughed, and in 1916, with a roar the first jet powered aircraft jet began its career.

With the jet stream exhaust roaring a few inches over his head the airplane rolled a few meters down the grassy surface and lofted into the sky!   Coanda managed to get it level and pointed it to pass over a small grove of trees at the end of the airstrip, (an achievement worthy of note given the primitive guidance and control mechanisms of the era.) There was only one trouble --- there was no way to throttle the beast! No one knows what speed it achieved before things literally began falling apart. First the Spad actually flew out of its wings, leaving them fluttering to the ground behind him! Then the varnished silk surface of the fuselage peeled off, taking the tail with it! The engine coughed and died as it ran out of fuel and the by now skeletal frame, with Coanda frozen to immobility, arrowed into the tops of the trees before settling down to fly no more!

Dazed, and understandably shaken, Coanda climbed down from the tree-top and, if he did not actually kiss the ground, he must certainly have felt like doing so!

As nearly as anyone could guess, his little jet had managed to reach a speed of perhaps 300 mph during its brief, 20 second flight. A Spad would have disintegrated at 175 mph. At 300 mph the disintegration was almost artistic; rather like a strip-tease dancer whose every move is choreographed. So Coanda held the unofficial world's speed record until World War II came along. But virtually no one knew about the affair. It was clearly too dangerous to fool with and Coanda himself quietly went back to his study of surface tension in liquids. The whole affair was quickly forgotten as merely one more bump in the road to winning a war. But it provides an interesting clue to the action of gravity if only somebody had paused to think about it.

Preliminaries out of the way, suppose we now get down to basics. Begin by noting that magnetism does not always attract. On occasion it repels, as when you try to push two north poles or two south poles together. Attraction is therefore variable with orientation. Pushing may also occur along with lateral pressure. Gravity, on the other hand, is a unidirectional force under all conditions. But despite these differences we can still make an argument for close kinship. So suppose we take another look at gravity. Seek out any possible flaws in the argument.

Begin by harking back to old Isaac Newton and his laws of action. "For every action there is an equal and opposite reaction." It sounds like an invariant dictum, but is it as invariant and absolute as it sounds?

Think of Coanda's little gadget. Now anchor one firmly to the surface of a massive metal plate. Next fire a burst of supersonic air directly into the jet intake. Add a little energy to the jet in the form of heat and record what happens. The result is startling. The jet motor is striving mightily to charge into the source of the wind!

In effect we have a force which attracts rather than repels when you push it! It isn't exact and there is more to it than I have said here, but it points out that there are some possibilities we ought to explore; some combinations which might solve the dilemma. Possibly we are too myopic in our thought processes

If this is not enough we might consider the behavior of an old time sailing ship in the wind.

Intuitively one might expect to achieve the maximum speed when the wind is squarely abaft the stern and blowing directly into the canvas, but this is not the case and the optimum result is achieved with a quartering wind!

None of this disproves Newton's law but it does point out that a little judicious tinkering can lead to interesting results. So how about taking another look at the Poynting-Robertson effect?

Solar radiation strikes a meter-sized sphere orbiting it in space at approximately Earth's distance from the sun. Given the omnidirectional reradiation of the received energy we may conclude that roughly 1/6 the radiation pressure received from the sun will be translated into acceleration away from the sun. The reradiation of the received radiation is omnidirectional and symmetrical so it largely cancels itself out save for that which is directed along the lead and following track of its orbit.

That which is expelled to the rear adds a minuscule dollop of momentum to the sphere. This increases its speed slightly so it climbs into a fractionally higher orbit. But that which is ejected in the direction it is moving more than counterbalances the other so it is now going too slow for the orbit it is in and it must fall slightly toward the sun in order to pick up the speed needed to exist in its new orbit.

This is despite the push imparted by the solar radiation so the same array of forces has actually increased in strength so the sphere is forced to repeat the cycle, continuing until the sphere either dissolves or simply plunges into the sun. In effect, *the radiation from the sun not only pushes it away, it also pulls it toward the sun, though in an unexpected way!*

True enough, if one considers all the forces acting in different directions and from different sources there is no argument with Newton, but a too facile intoning of the Newton mantra may prove dangerous. The system was in balance save for the solar radiation. In the crudest sense of the word then, solar push is over compensated for by a greater pull so Newton goes out the window! And where does this lead us?

Here is another bite to chew on: Newton merely points out that for every action there is an equal and opposite reaction. Even if we accept that the argument based on Poynting-Robertson may be somewhat flawed, if the picture is looked at as part of a whole, we still must consider that nowhere in Newton's dictum is any implication that the reaction must be immediate. It might even take millions of years before fulfillment. This harkens back to the section on kinetic energy in Part 1, where the bowling ball simply rotted away atop its perch so the potential energy evaporated. In short, the equal and opposite dictum ought not to be invoked too routinely. There may be factors we are overlooking. Bean counting is not without its hazards.

Start with a hypothesis and see where it leads us. Begin with the phote. I have already concluded that it is the sole occupant of the omniverse and the fundamental constituent of everything. Additionally, the phote must possess electromagnetic and mass/gravitational attributes. Lastly, electromagnetism and gravity must be different aspects of the same phote, with the sole distinction being that they behave differently according to circumstances.

Expand on this reasoning and see where it leads us. Suppose we were able to view a phote directly. What would we see? What would we not see? One thing we may be certain of; it would not be surrounded by an enveloping field. If it were, then the phote would cease being the simplest possible phenomenon in the universe. It requires energy to maintain a field and if a phote has energy to spare it would instantly become depleted and its field would vanish along with it to leave the energy stranded in a limbo which cannot exist.

But the presence of electromagnetism must be bi-polar or one phote must consist of a positive pole while another consists of a negative pole, and this has been pretty well foreclosed by repeatedly failed efforts to detect monopoles.

Now take this phote and aim it at an article --- in this case defining an article as any atom. Since every atom must ultimately be reducible to photes and particle physics is focused on the subatomic universe this temporary definition is useful though only provisionally accurate.

Postulate that the positive pole of the phote chances to approach at a positive pole at the aimed impact phote of the atom. Given this the outcome will see the alien phote ricocheting away from the target (bear in mind that being virtually massless the recoil will be next to instantaneous and moving at >C). As for the atom, the fractional mass of the phote creates a tiny shock wave which manifests itself as a rise in internal temperature along with recoil away from the source.

This is what is meant by the term 'radiation pressure'. In the simplest form we may visualize a simple solar cell. The sun's rays strike the cell and carom away as reflection, but the result is a rise in temperature which is bled off as electricity. It is not an exact analogy, but it is close.

Next suppose that it is the negative pole of the phote which slams into the positive pole of the phote at the focal point of the atom. Here we have absorption of the phote. This absorption causes a displacement of photes already there. A compressional wave is formed which is transmitted through the atom and a different phote is squeezed out on the far side. The kinetic recoil from this impels the atom to surge toward the source of the alien phote.

This is what we call 'gravity'. It is exactly what we see in a ram jet engine. Air is ingested at the front, compressed and heated and then thrust out from the rear, causing the machine to accelerate into the jet. We have the rays of the sun striking an atom as our model so I am merely copying nature when I point to the ram jet engine as an illustration of the way gravity works. And this was the unique contribution of Coanda to the theory of gravitation, but sadly, he died never realizing what he had done --- but no one else did either.

Do I believe that this is the solution to the problem of gravity?

Not really. There are many experiments to be done, and this is one area where bean counting is essential. So do not hold my feet to the fire on this note. All I have done it point to an utterly unexplored alternative to the problem; one largely manufactured by an obsessive determination to prove that subatomic particles each have a single defining characteristic. No one ever seems to ask *WHAT ELSE DO THEY DO*?

An alternative approach to the gravity question does not work in this analysis but is worth considering. Postulate that light *does* behave either as a wave or a particle. In this case light might enter an atom as a massless wave but this passes through the atom as a standing wave to expel an energy particle out the back side --- a particle which is essentially a phote and thus beyond the reach of our instruments. In effect, the wave enters into the atom with only an extremely tiny kinetic energy but exits on the far side with considerable energy. Here it becomes just another manifestation of the Poynting-Robertson effect? Is this not also a solution to the question of gravity? Is it not also in defiance of orthodoxy and a fresh apparition of the bean counter mentality? It certainly seems likely and avoids the paradox where a push results in a pull.

I cannot be wholly certain here, but it is evident that a myopic focus on only the alternatives --- gravitons or warping of space --- are not the only logical ones, and until all alternatives are explored it is premature to confine our theorization to any given one.

On this note I return to the step by step itemization of the process whereby our universe came to be. But I have not abandoned the problem of gravitation entirely. I shall return to it in Part 4, where I tackle the question of atoms and how they work --- not arguing against the current models, just filling in on a few places which have not been addressed to date.

Step VIII. The process of accretion is inescapable. As each of the proto-galaxies began to take shape they were subject to their own turbulence. As a crude analogy you might think of the motion in a basin of water as the drain is opened. Within a nascent galaxy these convergent forces are inevitably admixed and jostling against similar forces arising throughout the system. Individual stars are springing to life everywhere the galaxies, which had been little more than blobs of gas a few millions of years earlier, are busily taking shape.

Returning to the matter of an approximate time line between the first moment of the initial explosion and the emergence of stars, we may arrive at a probable minimum of 7 billion years. It may have been longer, but that would depend on circumstances beyond our calculations; such as varying densities of the local ylem, the number of eruptions in the chain explosions, the time involved in consolidating the chain into a solitary line, intensities of the e.m.f. emissions, etc. For any one factor I might be able to come up with a reasonable estimate, but when combined is a markedly ambivalent matter.

Broken down, the initial explosion would eject the photes out at immediately sub-c speeds, but the smoke ring would be a different matter. The circulation within the torus would be moving at >C speed while the forward motion of the torus itself would be measured in perhaps a few hundred km/sec as it arrowed through the ylem.

This would be implied by the resistance of the ylem piling up in the path of the torus and compacting to create its secondary blowout. It is difficult to imagine this happening quickly. Moving at 1,000, km/sec it would require 300,000 years to travel a single light year and at least 5,000,000 years to build up a bow wave dense enough to provoke a secondary detonation. Assume four successive eruptions in the chain and we have just spent a minimum of 2 billion years.

The time from the last eruption in the chain to the definitive consolidation of the torus was probably fairly short, but not insignificant since it had to involve systemic organization across a likely dimensional span of roughly a billion l.y. by this time. Total time now sitting at 3+ billion years.

Consolidation of the galactic clouds throughout the universe would be quite a slow process which must have taken as much as 4 billion years with the appearance of the first stars relatively quickly, perhaps on the order of two to three hundred million years.

Adding it all up, it is difficult to imagine this process taking fewer than 8 billion years and may have taken considerably more than this. But what we .have here is a minimum, not one minute of which is contemplated by orthodox dogma, which currently points to the oldest galaxies as dating back some 17 billion years, or 9 billion, or 12 billion, depending on which authority you accept. Add the 8 billion year figure I arrived at to each of these more orthodox figures and we arrive at an assortment of 15, 18, or 23 billion years as the minimum age of the universe.

Anticipating the discussion in Book 3, we know the solar system is some 4.5 billion years old, which allows only another 3.5 billion years for the population 2 stars to develop and evolve to a point where the population 1 stars begin to appear in significant numbers. There are questions here and arguments over details are entirely possible, but I believe the thinness of the time line forecloses the 9 billion year figure. The same may perhaps be said of the 12 billion age.

The 17 billion year age is feasible if we restrict it to the Hubble value for the remotest galaxies, but here again I must emphasize that the Hubble Constant is a fragile instrument for determination of the age of the universe, and even if it were measuring what astronomers believe it is, recent evidence argues that the rate of expansion has been steadily increasing over the aeons, which would make the 17 billion figure far too low.

In an effort to explain why the expansion speed of the universe has been increasing despite the expected slowing of its speed due to universal gravitation the theory boys invented a whole new fundamental force; one which they dub the '*cosmic repulsion factor*'. Given this, it is becoming increasingly clear that our poor, abused *empty* space has been utterly overwhelmed by the fantasies of our astrophysical theorists. Given a smoke ring universe which steadily expands and thins out as it encounters resistance by the ylem it is obvious that the whole 'cosmic repulsion factor' bit is simply silly.

It is sheer nonsense, but the intellectual gyrations performed to discredit the estimates are fascinating. I am personally inclined to propose a date more on the order of 35 billion years but this is merely a suspicion and has little save a few feeble hints to back it up.

So it all boils down to agreeing that our universe is getting along in years, but no one has agreed on a birth date. Only one thing is certain, it is not a 'snap, crackle, pop,' affair where an entire universe erupts, rather like an exploding cereal box out of nowhere virtually fully developed and ready to start doing business. Galaxies laden with stars are laid out in all their finery before our eyes through the Hubble telescope. It is much too complex --- *messy* might be a better word --- to be treated so simplistically.

My model takes a load off the shoulders of those who argue for a short existence of the universe where it all emerges from the darkness in 'one swell foop' (with thanks to Professor Spooner for his highly entertaining lectures). Much time and energy went into constructing our universe and enticing as the 'Big Bang, snap, crackle, pop' approach may seem at first glance, it simply does not work.

Even the computer gurus find their work cut out for them. Some pretty convincing equations argue that the relaxation time for globular clusters which appear to be escaping the hive exceeds the presumed age of the universe while other evidence acquired from the Hubble telescope and other arrays point to an age somewhere around 17 billion years... which just happens to be about one declared age of the universe (though a few crusty die-hards continue to argue for a more conservative 14 billion years, give or take a billion or so years; just as a safety precaution in case something else turns up.

Oddly enough, there is a distinctly unsettling alternative solution to this picture, providing anyone wants to settle for the 9 billion year mark, i.e. I suppose we might postulate the existence of another complete universe in a new, clandestine dimension apart from the 11 given here and argue that this other dimension became so overweight from its burden of stars and galaxies that it ruptured along a broad front and vomited out our universe as a fully formed, functioning universe. In that fashion we might argue that all the 'busy work' was completed before that universe entered into this universe to become 'our universe'.

Do I believe this? Not for an instant. But I am reasonably certain that some dedicated nuts out there will dance with glee at the prospect of roping in the suckers by claiming intimate knowledge of this other universe as derived from psychic auras sent out by the benevolent 'elder masters' of the great beyond, or some nonsensical variation thereof.

The fact that I scoff at an idea by no means I have disproved it, so please bear this in mind. Sorry if I keep reverting to this theme but I have noticed that most people listen only long enough to hear a personal bugaboo word or phrase then their mind clamps into lockdown and they do not hear a word you say after that. This peculiarity is not unique to any class of people, but oddly enough it appears to be more common among politicians, lawyers and professors than any other group save illiterates, The art of listening seems to be rather rare and one must belabor certain ideas in the (probably futile) hope than some of it will percolate through.

As long as I am digressing I may as well make it a good one. Ever hear of William of Ockham? Probably not, He was a Scholastic monk who set about writing analyses of philosophical writings and flourished around the start of the 14th Century. Today he is chiefly remembered as the author of "Ockham's razor", a dictum which seems to have been widely discarded over the past century. Succinctly, it takes dead aim at the all too common habit of introducing needless complications into the explanations of this or that phenomenon, and thus of lapses into obscurantism in hopes of concealing the fundamental weakness of your argument. In this Ockham was probably taking his own aim at a chap known as Peter Lombard (otherwise designated as 'The Labyrinth'), another medieval churchly philosopher justly famed for his habit of circular reasoning where he argued around and around until he was thoroughly confused, as were everyone else. He then escaped with a "Therefore!..." which mirrored the conclusion he had decided on in advance. For his part Ockham proposed that whenever two or more solutions are possible you should always select the simplest and forget the others. All in all it is a useful rule, but like so many rules it is not without risk. Since it is so widely ignored by modern cosmological practitioners it ought to be redefined for our brave new world, even though it may not be entirely correct.

Over the centuries, as problems have arisen they have been solved one way or another. In the process virtually all the easy solutions have been discovered, either by rational consideration or by accident through trial and error. Mankind has long since run out of the easy solutions and graduated to more complex problems, but the power brokers in all professions, including academia, are too busy flaunting their authority as savants to realize that the old rules no longer work and everywhere in the world failure of institutions has become endemic and a sense of hopelessness and unrest prevails.

The society fragments, with one party looking backward to the 'good old days', which never were while the opposing party is looking forward to some fantasy future which never can be. Neither faction shows much intelligence though both exhibit stunning arrogance in shilling for their pet fantasies.

This said, I now dismount from my high horse and return to cosmology. I suspect those who have persisted in reading my dissertation will be heaving sighs of relief at me quitting my digression, and I cannot say that I blame them, however they are a legitimate element in explaining how concepts and ideas, some dating back many centuries, have persisted as a substratum of our present outlook on things and continue to warp them to our detriment. Only if we recognize these cultural and theological habits for what they are can we hope to arrive at valid conclusions to our problems.

At long last and many digressions I have completed Parts I, II and III of my cosmology. Parts IV and V remain to be written, but it is only too likely they will never be written. I am into my 87th year aboard this little globe we call Earth so my time is limited. My memory, which used to be a source of pride for my ability to read it, understand it, and quote whole pages verbatim is now a fading relic and I must repeatedly retrace my logic 4 or 5 times to insure it is genuinely logical.

For the most part I believe I have succeeded, but the same sense of logic insists I must have failed on occasion. Even despite this I feel I have for the first time in history successfully laid out a useful framework for cosmology. It provides a no-nonsense overall schema where findings and theories can be hung while at the same time possessing enough flexibility to accommodate discordant data without having to resort to arcane, mystical postulations. If I have done this then I am content. If I have failed it will not be for any lack of trying.

XVII

COKE AND CODA*

With apologies to soft drink companies

I begin this last chapter with a definite sense of chagrin. Over the pages of this manuscript I have repeatedly alluded to globular clusters, always with a sense of doubt. They are utterly lacking a defined role in our galaxy so ordinary logic says they must have originated somewhere else and just drifted in as inclusions to the galaxy. But I was reluctant to focus on this alternative since other alternatives seemed possible, i.e., they might conceivably be relics of hyper stars which simply shed huge shells in a series of ordinary nova-sized pulses during the process of transitioning into localized clusters. This did not seem likely, but lacking information to the contrary I had to concede possibility. Or perhaps something else wholly unexpected might account for them. I did not know how this might have happened, but lacking data to the contrary there was always the possibility that we are merely fishing for some way to appear smart while the answer is hiding in plain sight right under our noses and I am too blind to see it.   I continued dithering over the problem, always without resolution, until the December, 2012 issue of Scientific American arrived in the mail. And there, to my wondering eyes did appear, not a jolly man in a red suit driving a reindeer powered sled, but a brief article by Dr. Anna Frebel pointing to the existence of aboriginal billions of mini-galaxies, too small to be picked out at great distances by telescopes but of the right size to develop into globular clusters susceptible of being swept up by the gravitation of larger galaxies. A close examination of Dr. Frebel's model proved pretty convincing and left me more embarrassed than ever when I realized I had not properly pursued my own chain of reasoning.

Consider the logic attending the chain of follow-forward detonations as the forming universe drifts through the surrounding ylem. The initial eruption occurs when the ylem concentration reached criticality but it cannot arrive at this point in isolation. Instead, it will be increasing in size while its speed of transit through the ambient ylem will be slowing.

My reasoning had taken me this far, but unfortunately I allowed my logic to stop there. Dr. Frebel's hypothesis prompted me to pursue the matter to a greater depth. Yes, the swirling vortex acquires new mass as it co-opts ylem from the local medium. Yes, it proceeds through the encompassing ylem without disturbing most, or even much, of the ambient ylem. In this it is not unlike the passage of a smoke ring through our atmosphere, i.e., there is ample empty space between the smoke particles and the local atmospheric molecules, but there will still be occasional photes (or molecules) which will be co-opted into the torus.

But this is not the end of the matter. A percentage of the ylem (or atmosphere) will be pushed ahead of the developing torus as a bow wave. This bow wave adds to the density of the ylem through which the torus is moving, thereby building up to a second, but smaller detonation. It is smaller simply because there is not as massive a density accumulation to draw from. Consider the sequence here. The initial explosion derives from a slow accumulation of photes drawn from a titanic pool of photes. This drains much of the pool and the developing torus moves toward out of the now pretty much depleted region toward a lesser density area which is not up to criticality. The bow wave from the initial torus provides an added supply of photes in a critical area so it reaches criticality and does its own exploding, but it is a smaller blast than the first because there is less seed material to work with.

Analogies to this may be found both in nature and in human affairs. Put another log on a fire and you get a hotter fire. Pus more gunpowder in the barrel of a cannon and you get a louder boom. Add more U-235 to a bomb and you get a bigger bang.

The initial explosion, which we may term the "mini-big bang", is the parent of the six potential universes. These in turn (assuming the potential six are all realized) created the first generation of hyper-stars --- which were the parents of the galaxies by providing an overwhelming gravitational nexus which sucked in the dwarf galaxies manufactured by the lesser eruptions along the way.

The Frebel model fitted in so perfectly with my own that I could not help but feel skeptical. It was simply too good to be true. So I decided to seek for possible flaws in my reasoning or tidbits of firm data which might be contradictory.

Finding none, I decided to fit it into the emerging schematic of this cosmology along with odds and ends of sustaining logic and follow that up with a couple of wildly unexpected extrapolations. Our Milky Way is host to more than a hundred globular clusters, with the smallest numbering 60,000 or more stars, all of them of similar age and composition. All are population 2 stars (with possibly an occasional 'guest' star which happened to venture too near the effective gravitation of the cluster or is perhaps in independent orbit about our galaxy and thus a mere passer-by. I do not know how accurate they were, and have not bothered to verify them by extended search, but on two or three occasions over the past half century astronomers have argued that the ages of some globular clusters appear to be greater than the life of the universe. This, of course, is nonsense in any frame of reference and I have consistently argued that the Hubble Constant is incorrect *ab ovum. But I am not critical of the authors of the observations*! They were merely pointing to an anomaly which indicated there was an error somewhere along the way and proved that a correction was needed which is perfectly legitimate science and happens to be pertinent in terms of a relationship to the age of our galaxy and thus in verifying Frebel's hypothesis.

The existence of the independent magnetic flux in intergalactic space raises several highly interesting possibilities. If condensed to conform to the mean dimensions of a dwarf (or mini) galaxy they might well produce 60,000 stars, all of which would bear the metal poor stamp of population 2 stars. If correct this might be taken to argue that the age of galactic creation is not over and done with --- which is not even hinted at in existing theory. It further suggests the likelihood that occasional dwarf galaxies (read 'globular clusters) may wind up merging with our galaxy over the next few billion years, joining the Magellanic clouds, M-31, etc. It further suggests that our universe is still incomplete and that continued slow growth will occur.

In the 1950's and 60's a trio of British theoreticians, Hoyle, Bondi and Gold, advanced what came to be known as what they called the "*perfect cosmological principle*', where the broad picture of the universe perpetually stays the same. It is not a mirror image sameness but it does argue that even as the universe expands its mean density remains static, i.e. as each galaxy slides beyond the reach of telescopic perception a like mass is created out in the void of interstellar space. This is not a case where new galaxies, or even individual stars pop up out of nowhere. It is merely a case where here and there a replacement hydrogen flickers into existence.

According to Hoyle and his associates this implies the addition of a solitary hydrogen atom per century per cubic kilometer of space in observable space. By itself this would be virtually impossible to detect, but the addition of so much mass over the billions of years gone by would result in new stars flaming to life within galaxies as well as interstellar space, or so the analysis went.

No evidence was found to support their hypothesis, nor were we able to detect the abrupt appearance of alien intruders in any of the neutrino detection apparatuses scattered about the world.

Lacking any empiric support a reluctant trio of astronomers abandoned their steady state theory in the late 1970's, where it has lain peacefully ever since. The Frebel hypothesis, when seconded by the electromagnetic field and my model for the birth of universes, provides an opening for a quasi-resurrection of the steady state theory. Other consequences were advanced from time to time, but there was never any hint of proof of the 'steady state'. On closer scrutiny however, it turns out there were logical flaws in the initial reasoning. To begin with, Hoyle, Bondi and Gold (HBG) calculations were based on the assumption that the introduction of replenishment matter is random throughout space. There is absolutely no reason to advance such a proposition and the presence of independent electromagnetic fields in intergalactic space may as easily be interpreted as indicating a preference of dark matter (ylem) to be co-opted into the torus if it chances to pass through neutral gravity regions within a universe. It need not be a strong preference, but it might be enough to pick up a random fleck of ylem from time to time --- which would more than provide enough new matter to accommodate the HBG model.

An alternative to this is more fascinating. Humankind appears to have a remarkably strong sense of pessimism. We were born with a sense of entropy and have since enshrined it as a dogma of faith. Everything runs down and erodes with time. The universe was manufactured in an instant by a word from on high and after a remarkably brief but volatile birthing process settled down on an inelastic path toward its ultimate demise, decaying from a golden age to a silver age and thence to a bronze an iron age in steps toward our devolution. Our species mythology declares us to be the supreme summit of evolution, with nowhere else to go but down. Things are born, grow to maturity and thereafter settle down to decay back into the clay from which we were fashioned.

But the presence of these electromagnetic fields and Frebel's dwarf galaxies raises the question of whether our torus may still be wending its way through the ylem and generating mini-bangs. If correct, our universe is still in its growth stage, not only in physical size but in terms of the acquisition of matter.

The prospect is interesting however I doubt whether the question will be resolved any time soon.

Almost as an aside here, please note that this puts the quietus to any prospect of using the relaxation age of globular clusters to offer any assistance in determining the age of our universe. The earliest clusters may long since have dissolved and merged entirely with their larger sister galaxies. Others may have relaxed and dispersed into empty regions between galaxies while the bright clusters we see in our own galaxy may be new arrivals freshly manufactured by the last mini-bang.

In the interim, while I would like to claim credit for the reasoning here I must cede the honor to Dr. Frebel. She richly merits it, the more so since both she and I were actively pondering the mystery of globular cluster origins and dynamics. She caught it while I missed it entirely.

POST CODA

Finally! After 90 years of learning and innumerable missteps, followed by 6 years of writing, I take my leave of the origins of the universe. I can at least promise that the next two parts, I,e, the lives of stars and questions anent the solar system will not be so convoluted. Here I have had to break new ground and generate a new way of approaching the problems. In the next two parts I will be venturing onto well-trodden pathways and mainly touching up on a few points of interest which appear to have been skipped or perhaps merely glossed over any practitioners within the field.

GWH --- December 23, 2015.